# The Life Story of an Infrared Telescope

More information about this series at http://www.springer.com/series/4097

John K. Davies

# The Life Story of an Infrared Telescope

John K. Davies
Astronomy Technology Centre
Edinburgh
United Kingdom
jkd@roe.ac.uk

Springer Praxis Books
ISBN 978-3-319-23578-3 ISBN 978-3-319-23579-0 (eBook)
DOI 10.1007/978-3-319-23579-0

Library of Congress Control Number: 2015955731

Springer Cham Heidelberg New York Dordrecht London


Cover Illustration: This fisheye lens photograph shows the UKIRT Wide Field Camera mounted above the primary mirror of UKIRT. The dome slit is open and the University of Hawaii 2.2 m telescope dome is just visible along the lower edge (Photo Paul Hirst)

Printed on acid-free paper

Springer International Publishing AG Switzerland is part of Springer Science+Business Media (www.springer.com)

*To everyone who contributed to UKIRT and who, in doing so, made this a story worth telling.*

# Preface

This is a story I have wanted to tell for a number of years, it was in my mind while I was still at UKIRT and the urge grew stronger once I had left. The attempt by the Royal Observatory Edinburgh librarian Karen Moran to organise a history of the observatory fanned the flames and the gathering of a number of key UKIRT people (if I may call them such) at the 30th anniversary conference in 2009 created an opportunity to gather considerable raw material. However it was the sudden, tragic, death of my good friend and UKIRT stalwart Tim Hawarden which provided the trigger to stop procrastinating and get on with the job. Once I had started the task turned out to be incredibly easy to continue, for I did not so much write this story as assemble it. The experience was like drinking from a hosepipe: simple e-mails to people I had not seen for many years, or even never met, produced a flood of material, almost all fascinating, some quite unpublishable, and more photographs than I could ever use.

Why should this be so? A telescope is a machine with an eye of glass, a skeleton of steel and muscles of copper and plastic. Its brain is a computer. It should not have a soul but, somehow, UKIRT does. Almost all of the people who have worked there, or even visited to use it, have an affection for this machine which transcends common sense. They almost love it. There is just something about UKIRT, or about the way it was run, which generated an ésprit de corps which survived long after the original builders and operators had left. So this is their story, or perhaps I should say our story, because I too fell under the spell of this machine.

I hope you enjoy reading about this adventure. If you were part of it, then I trust that I have done you justice; if you were never part of the UKIRT family then I hope that it conveys some of the excitement of working in this very special place.

Edinburgh, UK John K. Davies

# A Note on Units

As scientists, astronomers usually work in the SI (metric) system, but the United States and much of the United Kingdom still use many traditional imperial measurements in day-to-day life. This creates an interesting melange of terminology in which it is not uncommon to speak of a 3.8 m telescope on a 14,000 ft high mountain, or an 88 in. telescope at an altitude of 4000 m. In a transatlantic and cross-cultural adventure like this, it is difficult to know in which system to work. So rather than adopt one system and systematically translate into the other, this story is written in whatever vernacular the people of that time, and in that place, would most probably have felt comfortable. Conversions, rounded to a sensible number of digits, are indicated in brackets when a clear understanding of the dimension is important.

# Acknowledgements

Much of what appears here is only a lightly edited compilation of the many e-mails received and the interviews I have conducted. Specific quotes are in *italics*, but often the paragraphs in which they are embedded are largely written by the people involved in the incidents described. This material has been supplemented by extracts from newsletters, magazines and a few official papers. I owe all of my correspondents an incalculable debt, for it was their contributions that made this possible. They include:

Alan Bridger, Alan Pickup, Alex McLachlan, Alf Neild, Alistair Glasse, Andy Adamson, Andy Lawrence, Andy Longmore, Bill Parker, Bill Zealey, Bob Joseph, Charlie Richardson, Chas Cavedoni, Chris Davis, Chris Impey, Colin Humphries, Colin Vincent, Dale Cruikshank, Dave Robertson, Dolores (Walther) Coulson, Doug Simons, Doug Whittet, Frances Hawarden, Frossie Economou, Gareth Wynn-Williams, Gary Davis, Gillian Wright, Gwen Biggert (MKSS), Harry Atkinson, Helen Walker, Ian Bryson, Ian McLean, Ian Robson, Ian Sheffield, Jim Hough, Joel Aycock, John Clark, John McCarthy, John Morris (MKSS), John Peacock, John Rayner, John Womersley, Karl Glazebrook, Kent Tsutsui, Kevin Krisciunas, Liz Sim, Malcolm Currie, Malcolm Longair, Malcolm Stewart, Marc Seigar, Maren Purves, Mark Casali, Mark McCaughrean, Martin Ward, Martin Wells, Mathew Rippa, Matt Mountain, Mike Wagner, Nick Rees, Olga Kuhn, Omar Almaini, Pat Roche, Paul Hirst, Peredur Williams, Peter Forster, Philip Best, Phil Daly, Phil James, Phil Puxley, Ray Wolstencroft, Richard Ellis, Richard Jameson, Richard Wade, Roger Clowes, Rich Isaacman, Ron Beetles, Ron Koehler (MKSS), Sandy Leggett, Simon Dye, Sluing Yee Roth (Alfred Yee & Associates, Honolulu), Steven Beard, Steve Beckwith, Steve Warren, Stuart Ryder, Susan Beattie, Suzie Ramsay, Terry Lee, Terry Purkins, Tim Carroll, Tim Jenness, Thor Wold, Tom Geballe, Tom Kerr, Wanda Miller (Observa-Dome Laboratories Inc).

I am also grateful to Jim Gallagher and Winnie Schafer (STFC records office) who granted me access to some of the remaining UKIRT records of the early period, to Roger Griffin who had preserved some of the very early steering committee papers and other correspondence for posterity, and to John Jefferies (IfA) for access to his unpublished manuscript on the development of Mauna Kea.

Considerable use was made of articles published in the UKIRT Newsletter and in the ROE Bulletin, an internal staff newsletter published in the 1970s and 1980s, and I am grateful to STFC for permission to use this material.

Tom Kerr's blog "A Pacific View" kept me up to date with developments after about 2009 without me having to ask too many questions.

Finally, Tim Hawarden's talk at the UKIRT 30th anniversary conference formed the basis of Chap. 11. This whole story would have been so much better if he had been able to contribute directly.

As always, ultimately as author I am responsible for any errors, especially if I have misunderstood what I was told.

# Contents

# Chapter 1
# Conception

The father of the telescope we today know as UKIRT was Professor Jim Ring of the Imperial College of Science and Technology, London. He was an experimental physicist who graduated from Manchester University in the 1950s, lectured there for a time, moved to the University of Hull in 1961 and then to London in 1967 where he set up an infrared astronomy group. Jim Ring was interested both in both astronomical instruments and in telescope design and wanted to explore the limits of telescopes as the diameters of their primary mirrors was increased. In particular he sought to challenge the canonical cost-to-diameter relationship for telescope construction which said that, because of the cost of steel and concrete, the budget for a telescope of traditional design increased as about the third power of the primary mirror diameter. At a time when one arcsecond seeing was assumed to be the limit achievable from the ground, he wanted to know how big a mirror with acceptable image quality could be made. He believed that by replacing a conventional primary mirror, i.e. one with a diameter to thickness ratio of about 6 to 1, with a thinner one supported by air filled pads the mass of the mirror and its supporting cell could be significantly reduced. In turn, the mass of steel and concrete needed to support the optics, and hence the cost, would be lower and this virtuous circle would make larger telescopes more affordable.

Preliminary calculations of a thin primary mirror supported on an axial support system of 80 pads positioned along three radii were performed by John Long, a mechanical engineer in Ring's group at Imperial College. These showed that with such a support system the root-mean-square deformations of the primary mirror, and therefore degradation of the telescope image, could be limited to a fraction of a micron over the entire surface of the mirror, giving an image that would be diffraction limited at mid-infrared wavelengths. Since the internal noise in the lead sulphide (PbS) detectors used for infrared astronomy at the time was greater than the combined thermal noise from the sky and the telescope, photometric measurements taken through apertures of several arcseconds diameter did not require particularly good image quality. So a telescope of this design, often referred to as a "flux collector" or less flatteringly as a "light bucket", would be quite



J.K. Davies, *The Life Story of an Infrared Telescope*, Springer Praxis Books,
DOI 10.1007/978-3-319-23579-0_1

acceptable for observations that did not need images limited by the atmospheric conditions.

In 1966 Sir Harrie Massey and Jim Ring organized a Royal Society discussion meeting on the newly developing field of infrared astronomy. The proceedings of the meeting, which were not published until 1968, included a paper by Peter B. Fellgett entitled "Large Flux Collectors for IR Astronomy", although a popular version of essentially the same paper had already appeared in a magazine called Science Journal. Fellgett, who was by then the Professor of Cybernetics at the University of Reading, had written a PhD on infrared spectroscopy at Cambridge in 1949 and subsequently worked at Cambridge, the Lick Observatory and at the Royal Observatory, Edinburgh (ROE) during the 1950s and 1960s. Fellgett's paper described low-cost, lightweight infrared flux collectors of relatively poor image quality with a typical upper size limit of 120 inches (3.05 m). The paper remarked that '*2 or 3 such units located in Britain and Europe (one at a high altitude station) offer great discovery potential*'.

By 1967 thinking along these lines had advanced to the point that a proposal to build a medium-sized prototype of an IR Flux Collector (IRFC) was made by Ring to the UK's Science Research Council (SRC) and was approved in 1969. The instrument was originally conceived as a portable telescope housed in a roll-off shed that could be moved to various locations to conduct site testing for a larger, as yet unfunded project, but was eventually turned into a more permanent facility located at Izana on the island of Tenerife. The 1.5 m telescope saw first light in April 1972 and was used by UK astronomers, including those who had returned from undertaking PhD studies in the USA, to develop their skills in building instruments and making observations in the infrared. Crucially, the IRFC also indicated that the low-cost flux collector design concept was sound and in 1973 design studies for a scaled up 4 m version were commissioned from the firms of Grubb Parsons in Newcastle and Dunford Hadfields in Sheffield, both of whom had been involved with the 1.5 m IRFC. These studies proved encouraging and Ring proposed to SRC that they should follow up this prototype with either a 3.8 m flux collector situated at Izana or a 3 m version to be built on Mauna Kea, a dormant volcano on the Big Island of Hawaii.[1] Gillian Wright, then a PhD student at Imperial College who was making infrared observations of interacting galaxies under the supervision of Bob Joseph, remembers that it was Ring who pushed the idea that the UK should build these telescopes, and who spent his time at committees and reviews and the like. While admitting that her detailed memories are dim, she says that '*The things I remember were the impression of the passion and enthusiasm of it all*'.

[1] The state of Hawaii is a chain of islands and while the majority of the population and the centre of political and business life are in Honolulu, on the Island of Oahu, the largest island is Hawaii itself. Hawaii is known locally, and for obvious reasons, as "the Big Island".

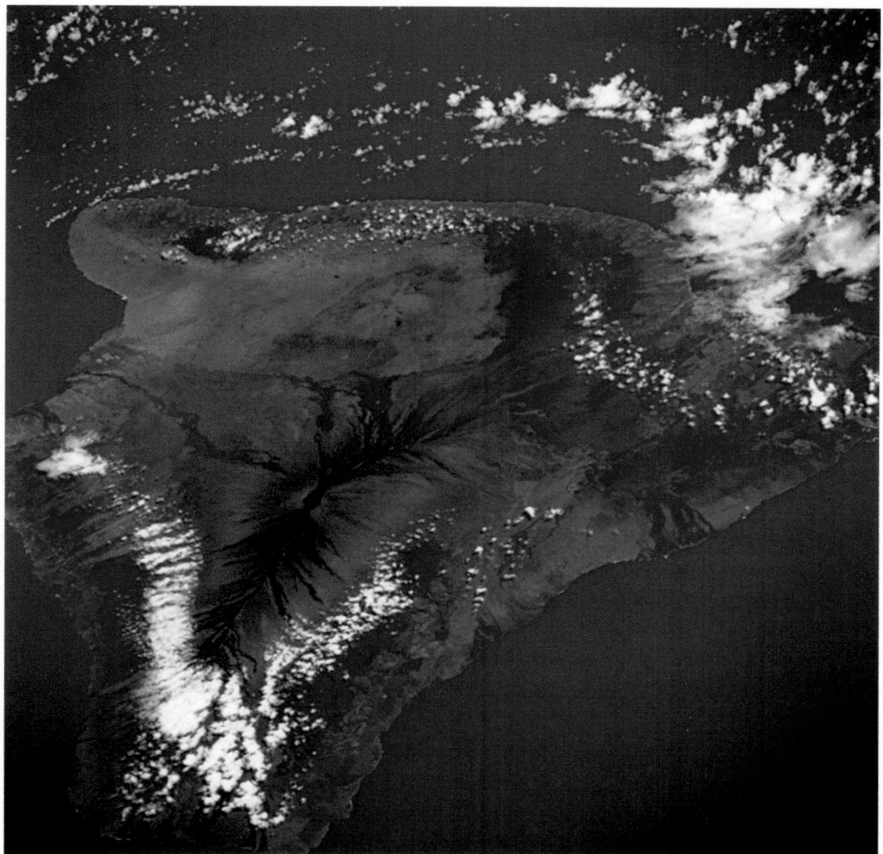

**Fig. 1.1** Hawaii from space. The town of Hilo is around the bay almost covered by clouds in the upper right. The summit of Mauna Kea is the small brown area surrounded by green pastureland near the centre of the island (Photo NASA STS-61-50-57)

The Hawaii option was of interest for two reasons, one scientific and one programmatic. The summit of Mauna Kea is 14,000 ft (4200 m) high, so the air above it is both thinner, and crucially, drier than is found at lower altitudes and so offered obvious scientific gains, indeed it was already the site of an 88 inch (2.2 m) telescope operated by the University of Hawaii who had been granted a 65 year lease of the summit area in 1968. On a practical level the UK had been drawing up plans for a suite of new optical telescopes for what it called a "Northern Hemisphere Observatory" or NHO for a number of years. Site testing of three sites in Italy, Almeria in mainland Spain and on Tenerife was planned and, while the Almeria programme fell foul of political issues (the Germans also hoped to build here), site testing in Italy and Tenerife was under way in 1972. By 1973 the Canary Islands and Hawaii were the clear front runners for the location of the national observatory and John Hutchinson of the SRC's Astronomy and Space Research (ASR) division had been discussing with John Jefferies, the then Director of the Institute for Astronomy (IfA) at the University of Hawaii, the possibility of building the NHO on Mauna Kea. If that plan came to fruition the NHO might incorporate the proposed IR Flux Collector as well. However, the NHO decision was yet to be made and the UK was actively pursuing options for it to be located closer to home so, in late 1973, Hutchinson wrote to Jefferies to explore the possibility of building the IR Flux Collector as an independent facility close to the existing telescopes along the summit ridge on Mauna Kea. Jefferies' reply was encouraging and the SRC's Astronomy II committee, chaired at the time by Professor Walter Stibbs, made the bold decision to go for the larger telescope and to site it in Hawaii. In January 1974 the 3.8 m IR Flux Collector proposal was accepted in principle by the SRC's Astronomy and Space Research Board who asked for a detailed submission to be prepared for final approval.

To finalise the proposal it then became urgent to establish the terms of any agreement to build on Mauna Kea, so Hutchinson arranged to visit Hawaii as soon as practicable. He was to be accompanied by Gordon Carpenter of the Royal Observatory Edinburgh. Carpenter, who was sometimes known by the nickname "Chippie", had relinquished his position as the head of the instrumentation division at the ROE to work full time on the flux collector project and found himself seconded, on a part-time basis at least, to Jim Ring's group at Imperial College. The visit to Hawaii was set for early March 1974 and, in preparation for the negotiations, the Astronomy Policy and Grants Committee prepared a detailed brief for the two men to take with them. This briefing included a guideline that 10 % of the observing time could be offered to the host institution in lieu of rental costs for the site. The visit went well, with the two men visiting the University of Hawaii on the 4th and 5th and agreeing, subject to the legal details being worked out, on a site 200 by 150 ft (61.5 by 46 m) along the summit ridge, near the University of Hawaii's (UH) 88 inch telescope and the then proposed Canada-France-Hawaii Telescope. Since it was en-route to the existing UH 88 inch telescope building the proposed site had the advantage of requiring no new infrastructure. The lease was expected to be for 25 years (after many delays it would end up being 27 years), extendable in 5-year increments, with the UH to be given the option of taking over the building, and perhaps the telescope itself, when the site was vacated.

In his unpublished manuscript about the setting up of the Mauna Kea observatories, John Jefferies wrote about the setting up of the agreement. *'We at the IfA had adopted the policy of asking for a guaranteed share of the use of every telescope as a sort of ground rent. Of course the land did not belong to the University, still less to the Institute for Astronomy, and the State could have insisted on a monetary payment for rent had they wished, but the issue simply never came up and I was certainly not going to raise it'. 'I suppose no one thought to question our appropriating to ourselves the effective right of ownership. Certainly it was the fact of receiving guaranteed telescope time in exchange for use of the land that underlay the growth of the Institute for Astronomy. The negotiations with the UK people were characteristic of the ease of our relationships – I said we needed 20% of the telescope time – they (I guess it was Gordon [Carpenter]) said that they were thinking of 10% after which 15% was suggested and immediately adopted. The exchange took less than a minute'*. In fact it was slightly more complicated, the SRC representatives' brief was to surrender 10 % of the time and they had no authority to agree to an increase in this figure. They merely agreed to take the proposal back to higher authority in the UK who did indeed accept the 15 % figure.

With the basis of an agreement in hand the proposal for the project went back to the ASR board, who on the 15th of March 1974 agreed that the case should go forward to the highest decision-making level of the UK scientific process, the Council of the SRC. The Council considered the proposal at its meeting on the 17th of April and gave the go-ahead. The budget of the proposed telescope, excluding instrumentation, simulators to be built at ROE and instrument related work to be done in universities, was set at £2,500,000. Since this exceeded £1,000,000, the project had to be approved by the government's Department of Education and Science, but Stibbs and Ring were able to make the case and the project was fully approved on the 18th of June 1974. Jim Ring was appointed project scientist and, with the ROE put in charge of construction and commissioning, Gordon Carpenter was made project manager. The project was publicly announced by SRC in early August 1974.

# Chapter 2
# Design and Planning

## Design

Once the project was approved the Astronomy Policy and Grants Committee of SRC appointed a steering committee for the construction stage of the project, still known officially as the "3.8 m IR Flux Collector". This was chaired by Jim Ring with Robert "Bob" Stobie of ROE as secretary and included Richard Jameson, David Allen, Phil Marsden of Leeds, J. C. D. "Lou" Marsh, Mike Selby, Michael Smyth and Gareth Wynn-Williams. Roger Griffin, of Cambridge University, was asked to join the committee to represent the views of optical astronomers. Victor Clube was included in the original appointees but resigned after the first meeting and was replaced by ROE astronomer Terry Lee. Although the SRC records of the committee were destroyed in the 1990s the terms of reference survive. They were to:

1. Within the broad scientific objectives and financial and other constraints laid down by the Astronomy II committee, advise those (normally the project manager) with direct accounting responsibility for the construction of the project about the specification and other aspects of the project to attain the desired scientific ends;
2. Regularly consider progress reports and financial statements by the Project Manager and advise him accordingly; and, similarly, consider proposals to commit expenditure on items of value £5000 or more;
3. Advise the Astronomy II committee on progress periodically or whenever the broad framework laid down by the Committee is in question.
4. Maintain close contacts with those planning the Northern Hemisphere Observatory.

The first meeting of the steering committee was held on the 6th of June 1974 in London. Papers were presented by Gordon Carpenter and Jim Ring on the state of ongoing negotiations with the University of Hawaii and on the plans for the



J.K. Davies, *The Life Story of an Infrared Telescope*, Springer Praxis Books,
DOI 10.1007/978-3-319-23579-0_2

telescope and its building. The draft specifications of the telescope presented were as follows

| | |
|---|---|
| Primary diameter | 3.8 m, as near as possible to f/2.5 |
| Focal stations | f/9 –f/12 Cassegrain, ~f/16 Coudé |
| Image quality | 98 % Encircled energy in a 2.4 arcsec circle |
| Nod time | 2 seconds |
| Tracking | 5 arcsec per hour |
| Pointing | 30 arcsec rms |

Other key design considerations were for a thermally clean structure (so no complicated optical baffles were included), a lightweight structure with no central box to minimise costs, the ability to move quickly for nodding and chopping and a computer controlled system to avoid the need for a lot of analogue circuitry. The input of pointing information was to be via a keyboard, doing away with the need to dial up values on thumbwheels. Much of the technical support, and the use of a telex machine in Hilo, was expected to come via agreements with the University of Hawaii. The observing floor control room was specified for up to three people, as was the combined lounge/kitchen to be built downstairs.

One of the first actions of the new committee was to evaluate and propose amendments to those requirements and they recommended an increase in the payload of the instrument support system to 200 kg, anticipating the day when instruments outgrew the typical 25 kg of contemporary photometers (but quite failing to appreciate the size of instruments like CGS4 which would arrive almost two decades later). The other main change was to increase the enclosure diameter to 60 ft (18 m) to allow for the addition of a chopping secondary system on the top-end. That would provide an alternative to using a chopper in the telescope's focal plane as had been done at the 1.5 m prototype in Tenerife.

The decision to provide the option of a chopping secondary mirror is an interesting example of how improvements in technology happening in parallel with the development and approval of a large project can change the specification and impact the cost significantly. When the telescope was first considered, infrared detectors were such that observations short of about 3.5 μm were limited by intrinsic detector noise rather than by the thermal characteristics of the telescope structure. It was only for observations at wavelengths longer than about 5 μm where the thermal noise was the dominant problem, and experience showed that in that situation using a chopping secondary mirror gave better results than chopping in the focal plane. Chopping secondary mirrors had other advantages too, such as increased chopping range and offered possibilities of image stabilisation. However advances in detectors, for example the Indium Antimonide (InSb) semiconductor devices that were becoming available, were steadily moving the detector limited constraint to shorter and shorter wavelengths. The strong case put by Gareth Wynn-Williams and from astronomers in the US convinced the Steering Committee that a building that could not accommodate a chopping secondary presented an unacceptable risk to the potential future scientific exploitation of the telescope. Nonetheless, since the project design and cost was based on the smallest building that could accommodate the telescope; the inevitable consequence of a longer telescope was a larger, and more expensive, building.

The provision of the chopping secondary itself would take a while to resolve, but the issue was solved by an agreement that an f/35 chopping secondary would be bought from the instrumentation, rather than the telescope, budget. The size of the secondary mirror was also debated at some length. An undersized secondary prevents thermal emission from the telescope structure being seen by the infrared instruments but is of little value to optical astronomers for whom it simply wastes some of the light garnered by the main mirror. Bob Joseph, an American astronomer then working in London, is *'fairly certain the initial specification did not include an undersized secondary, because I remember rather extensive discussions with Jim Ring about this. He really thought of the telescope as a flux collector, rather than an optimized infrared telescope... It took some extensive pressure to get him to relent and accept an undersized secondary mirror'.*

As early as the first steering committee meeting it seems that attempts were already being made to talk up the idea of improving the mirror performance from the bare minimum needed for a flux collector. After the meeting, which he had been unable to attend, Roger Griffin wrote a letter to Jim Ring supporting the idea, already floated by others, of improving the mirror specification. He wrote *'It seems to me that every slight improvement that may be made in the figure will bring benefits in the way of improved potency of the telescope'* although he recognised that such an improvement might make the telescope *'too attractive to optical people who might then offer significant competition for observing time'*. Jim Ring replied supporting the idea of going for the best possible image specification provided that it was done within the other constraints. He went on to note that the only reason they could build a 150 inch (3.8 m) telescope for just over £1,000,000 was because they had relaxed the image specification somewhat. There was, he said, *'little point in polishing the mirror to a better figure than its support will allow or than the drive accuracy will allow us to use'*. In about August 1974 the agreed 3.8 m IRFC performance specification was sent to companies who might wish to tender for the contract.

The proposed optical and mechanical design was based on those primary mirror blanks that were immediately available at a reasonable cost. A Zerodur blank could not be available before May 1976 but Owens-Illinois had a 3.8 m CER-VIT blank in stock. This blank had been cracked but it was sawn in half to produce the very thin mirror demanded by the flux collector design. After that dramatic piece of surgery it weighed about 7 tonnes, about one third the weight of a traditional mirror of that size. Grubb Parsons of Newcastle upon Tyne were selected to grind and polish the optics on the basis of their recent experience producing the mirrors for the Anglo-Australian Telescope. Grubb Parsons were unsure of how accurately such a thin mirror could be figured, but in view of the recommendations of the steering committee, the contract contained a provision for continuing to figure the primary mirror if testing indicated that it was possible to reach a higher specification. If such further polishing were to prove feasible, then a price for any improvement could be agreed.

## Preparing the Mirror

Once the mirror blank had been delivered to Grubb Parsons in 1975, polishing operations began with the provision of an 80-pad axial support system, cutting out the 1 m diameter central hole and the grinding of the inner cylindrical edge surface around the hole. Next came grinding the cylindrical edge surface around the outer diameter and the grinding and figuring of the top surface of the mirror. Grubb Parsons also cemented three radial defining Invar pads in the inner bore of the mirror and 24 Invar pads around the outer diameter to provide contact with the counterweighted outer radial lever arms of the mirror cell. For the required paraboloidal shape to be reached, the volume of CER-VIT glass that had to be removed by grinding the top surface of the originally flat disc resulted in a depression nearly 10 cm deep at the centre of the mirror. The figuring of the primary mirror required the manufacture of supports for the polishing laps, which were attached to a programmable rotating mechanical drive, fitting the laps with new pitch surfaces for each polishing stage, and then operating the system with progressively finer abrasive material as the process continued. Each figuring cycle took 2–3 days after which the mirror was washed, dried and then moved across the floor of the optical shop to a test tower where the surface profile was measured. That was done by Hartmann (knife-edge) measurements at the mirror's centre of curvature, some 19 m above the mirror, and also by shearing interferometry so that the optical performance at each polishing stage could be evaluated. The total time between each testing cycle was 7–10 days, including the time that it took to analyse the optical test measurements. Before each new polishing cycle began the entire surface was carefully inspected to determine whether there were any air bubbles near the surface that would be penetrated by the next polishing stage. If any such bubbles were found they were drilled out to prevent tiny shards of glass from the broken bubble being dragged round by the polishing laps and scratching the polished surface.

The work on the mirror was performed in the optical polishing workshop of Grubb Parsons where, by a strange coincidence, the foreman was named Jim Ring. Normally the figuring operations were done during the working day and then stopped overnight before restarting the following morning. One night a fluorescent lamp high in the roof of the workshop, and immediately above the mirror, disintegrated and when the staff arrived in the morning the shattered remains of the 1.5 m fluorescent tube were lying on the mirror surface. Fortunately, the damage to the mirror was negligible; there was only one very small scratch on the surface where the largest fragment of the tube had impacted. To prevent any recurrence of this potentially disastrous accident all of the fluorescent lighting tubes in the optical workshop were covered with fine mesh screens.

In addition to figuring the telescope primary mirror, Grubb Parsons was also responsible for providing a 1 m diameter f/9 secondary mirror and a flat mirror for a coudé focus.

## Building the Telescope

The firm of Dunford Hadfields of Sheffield was chosen as the contractor for the telescope structure and its drives, all of which had to be designed to survive the effects of the frequent earth tremors expected on a volcanic island. The industrial project team included: Des Hickenson (Project Manager), Alan Deeley (design and production of control system hardware), Alex Gaymond (mechanical engineering), and Les Wilson (mechanical engineering). Design engineering was the responsibility of consultant Denis Walshaw. The telescope mounting was an English yoke type, with two columns carrying the north and south bearings resting on concrete piers extending down into the cinder cone on which the telescope would be built. The 20 tonnes, 12.2 m $\times$ 6.7 m, rectangular yoke that carried the telescope's open-tube framework was constructed as a welded steel box section that rotated about the polar axis which, at the latitude of Hawaii, would point at an angle of approximately 20° above the horizontal.

**Fig. 2.1** This model of the telescope features a cut-out photograph of Alex McLachlan which was pasted onto an original picture and then re-photographed. The picture appears in the 1974/75 ROE annual report

A consequence of the decision to use an English yoke mounting was that the telescope would be unable to point further north than about 60° declination because the northern pier would prevent the tube being moved low enough to see the polar regions. The northern limit would make it impossible to observe a number of interesting astronomical objects, but it was considered an acceptable penalty of opting for the cheapest possible telescope. Learning lessons from the IRFC in Tenerife, the declination bearings of the new telescope were designed to be stiffer than those of its predecessor to reduce shaking by the wind, but that effort would turn out to be only partially successful. The radial bearings for the rotational axis were supported by the two steel columns, the feet of which were tied by a braced under-frame. The gravitational load of the whole assembly, some 80 tonnes, was taken by a set of three ball bearing races resting on the concrete piers under each of the support columns. The telescope framework (a Serrurier truss system) carrying the primary mirror cell and the secondary top-end unit rotated within the yoke about the declination axis radial bearings.

## Altitude Worries

Even before either project had been approved, the protagonists of both the 3.8 m IRFC and the NHO were aware that a location on Mauna Kea would expose both staff and visitors to the challenges of working at high altitude, and that the risks needed to be fully understood before any decisions were made. So in October 1973 Gordon Carpenter, Professor Roderick Redman from Cambridge, who sat on the NHO Planning Committee, John Pope, a telescope engineer from the Royal Greenwich Observatory in Sussex, and John Hutchinson from the SRC visited the Royal Aircraft Establishment (RAE) at Farnborough.

The RAE, then the centre of the UK's aviation research expertise, had considerable experience of the effects of altitude on pilots and so was an obvious place to look for advice. The report of the visit includes discussions regarding the need for oxygen masks and introduced an interesting non-SI unit indicating that the effect of altitude is roughly equivalent to alcohol and that 'A man coming up to 14,000 ft from sea level would be 1–2 Martinis down'. The conclusion seems to have been that a 3-day period at an intermediate mid-level facility would be needed for acclimatisation and this conclusion was transmitted to the steering committee. However, Roger Griffin argued for a bedroom to be included in the building design since he felt that the proposed mid-level accommodation at a place known locally as Hale Pohaku (altitude 9200 ft, (2800 m)) was too low for it to be a suitable place to acclimatise. He also wrote a paper to the Northern Hemisphere Observatory Planning Committee in which he advocated building the residence for astronomers on the summit itself and allowing a two-night acclimatisation period.

Griffin had actual experience of the situation in Hawaii; he had hiked up the nearby mountain of Mauna Loa half a dozen times, staying for about a week or so in a National Park Service hut on the summit. He discovered that although he certainly felt the altitude upon arrival, the effect wore off altogether after 3 or 4 days. To prove the point on one occasion he took a picture of himself (with the self-timer on his camera) standing on his hands next to a notice indicating an altitude of 13,000 ft. He '*had 8 seconds to get stabilized in a handstand after pressing the button, and I did it at the first try! But when I showed the picture to someone, he thought it proved the reverse of what I thought it did!*' The whole issue would surface again a few years later when construction and commissioning of the telescope was set to begin.

## The Long Road to a Lease

Despite the enthusiasm of John Jefferies and the IfA for building telescopes, the summit of Mauna Kea was a state conservation district so before any commitment could be made to construction work it was necessary for the Institute for Astronomy to provide an Environmental Impact Statement on the likely effect of the present and planned facilities. Jefferies was aware of the potential delays that this might engender and in late November of 1974 he had met with the Governor of the State of Hawaii who had provided assurances of support from himself and most of his cabinet for astronomical development on Mauna Kea. Nonetheless, an impact statement was required and one was duly drafted in the spring of 1975. While noting the potential visual impact on the landscape and various related issues such as effects on the dirt road, soil, drainage etc, it concluded that the negative effects of both the UK Flux Collector and a similarly-sized NASA telescope, the InfraRed Telescope Facility (IRTF) would not be serious. More encouragingly, it also remarked that the socio-economic impacts (basically money and employment coming into the state) would be positive and long-term, if only "moderate". Permits for both the UK telescope and the IRTF were approved on the 29th of August 1975 and by October a sub-lease agreement between SRC and the University of Hawaii was in preparation. In it the SRC committed to design, gain approval for and then fabricate the telescope and its building on the summit. It further committed to paying a, still to be negotiated, share of the building and operational costs of new mid-level facilities to be built at Hale Pohaku and a share of the cost of maintaining the road from Hale Pohaku to the summit. The new mid-level facilities were to include a complex of office, laboratory and apartment buildings within which five apartments and two townhouses plus two laboratories were to be set aside for the use of the SRC. Until such facilities were built, SRC was empowered to build its own temporary structures for its contractors and staff. In return for its 15 % of observing time the University of Hawaii agreed to provide an access road, electrical power up to a maximum of 75 kW and a telephone connection to the Hawaiian telephone company network.

In fact, finalising the lease and some associated documents concerning the commitments on both sides took quite a long time. SRC were anxious to proceed with the project and did not want the whole process held up by detailed negotiations over the proposed facilities at mid-level which although "promised" by John

Jefferies had yet to be funded. In April 1975 Gordon Carpenter met John Jefferies and Ginger Plasch of the IfA and his notes indicate that he hoped the documents could be signed in a small ceremony to be held in London on the 30th of June or the 1st of July 1975. The notes did however caution that there was a 20 % chance of a delay if a further state agency, the Office of Environmental Quality Control, whose head seemed less sympathetic to the project, were to reject the Environmental Impact Statement and demand a new one. In fact no such ceremony took place because the 20 % chance came up.

Although SRC believed that an acceptable draft had been arranged by December of 1976, a major stumbling block turned out to be concerns from the Hawaii Department of Land and Natural Resources (DLNR) over the issue of how SRC could guarantee that the site could be restored to its natural state if the UK chose to leave and the IfA refused to take over the building. Astonishingly this issue rumbled on throughout 1977 with SRC becoming increasingly anxious to sign a deal, since by then construction had already started and they feared severe criticism from parliamentary watchdogs like the Public Accounts Committee for proceeding with the project before the sub-lease was signed. The University of Hawaii tried to break the deadlock by offering to restore the site on behalf of the SRC in return for certain assurances from the UK. Although the appropriate government agency, the Department of Education and Science, was willing in principle to make such undertaking, agreeing a suitable form of words was a slow process and by the time it was done the authorities in the DLNR had hardened their attitude. They returned to an earlier position that a University of Hawaii guarantee was not itself sufficient and were insisting on some kind of negotiable cash bond (a figure of $100,000 was mentioned) on which they could call should the UK default on its promises. To be fair this was not a slight on SRC itself, it was proposed that similar demands were to be placed on other facilities, but to have such an issue appear so late in the process was a major problem for the SRC who began to debate if intervention at a diplomatic level might be required. In the end the problem was solved by the State of Hawaii agreeing to accept a written guarantee in lieu of a monetary bond and the DLNR finally approved the sublease at its meeting on the 18th of November 1977.

Once again steps were started in the UK to obtain a letter from the appropriate government department. This took almost another year and the sublease, between the University of Hawaii and the SRC was eventually signed by the SRC chairman Sir Geoffrey Allen on the 21st of September 1978. At about the same time the underlying operational agreement, which detailed the practical arrangements between the SRC and the University of Hawaii, was signed by SRC's Harry Atkinson and the University's contracts office. Incredibly, by that time the telescope was already half built. Atkinson says *'Basically, we all wanted to get the telescope up and working – but things moved slowly with the bureaucracy in Honolulu and Hawaii. On the SRC side, things went faster than I've ever seen!'*.

In late 1974 Vincent Reddish, then a staff member at ROE but who was to be appointed ROE director from the 1st of October 1975, had estimated that, given its commitments to other national facilities such as the UK Schmidt Telescope in

Australia, the NHO and the COSMOS plate measuring machine, the ROE could provide at most 15–20 people for the Flux Collector project. Of these 7–10 might be committed to instrumentation projects with the rest being on the telescope construction team. In 1976 the SRC made ROE responsible for the support of the telescope once it was complete and, presumably knowing that this was coming, Reddish had already begun to put in place a management structure for the operational phase. His structure matched the operational model which had been adopted for the ROE managed UK Schmidt telescope, with what would eventually be called the "UKIRT Unit of the ROE" based in Edinburgh having overall management responsibility for the operation and whose head reported to Reddish as ROE director. This unit would in turn oversee a team led by an Hawaii-based "Officer-in-Charge" who would be on secondment from ROE. The structure was intended to keep control of the telescopes under the management in Edinburgh but it would, in due time, lead to tensions between the two ends of this chain.

To lead the ROE-based team Reddish identified Ramon (Ray) Wolstencroft, a British astronomer then working at the IfA in Honolulu. Ray had graduated from Cambridge and in 1962 had taken a junior astronomer position at the newly formed Kitt Peak National Observatory in Arizona. Having accepted a Fulbright grant to cover his travel to the USA, he later discovered that he was obliged to return to the UK after 3 years as a condition of that award. Ray took up a position at ROE in 1965 before becoming an associate professor at the University of Hawaii in 1968 at the invitation of John Jefferies. In 1975 Reddish visited Hawaii and encouraged Ray to return to ROE with the intention that, in due course, he would become Head of the UKIRT unit. It was implied that at some time Ray would return to Hawaii to lead the operations there. Ray remembers *'Flying over and going straight from the airport to somewhere in High Holborn then sitting in front of a panel of pundits and trying to stay awake to answer their questions'*. Ray was appointed a Senior Principal Scientific Officer at ROE from May 1975 although he did not physically arrive in Edinburgh until October. At the time a UKIRT Hawaii staff of about ten was envisaged, with only one or two of these being astronomers. The others would include five night assistants, a technical assistant and an administrative assistant, all overseen by an Officer-In-Charge. These estimates would soon rise dramatically as the realities of operating a telescope at 14,000 ft began to sink in. By about 1977, at which time "round the clock" operations were being considered, the estimated staffing levels had risen to about twenty, a figure which included five night assistants and a small team of technicians on call at the mid-level facility. Since posting ROE staff overseas was expensive it was hoped that the staff requirement could be minimised by having people double-up with, for example, technicians operating the telescope on shifts when their specific engineering skills were not required.

# Chapter 3
# The Project Advances

## Building a Team

During the project preparation phase in 1974 Gordon Carpenter, assisted by Janice Murray, was gathering together the people he would need to carry it through and, as things began to develop, this team started to expand. Gordon Adam was included in 1974 as was Alex McLachlan who became the project co-ordinator in October. Alex was not an astronomer but he had been trained to use the observatory's 36 inch telescope which, at the time, was used every clear night during the winter. In the autumn of 1976 Colin Humphries was asked by Vincent Reddish, by then Director of the ROE, to provide extra support to the UKIRT project because the number of ROE people working on it at that time was so small and, in any case, the project would need extra manpower during the construction phase. Colin was a research physicist who had been working on the results from the European TD-1A satellite's ultraviolet telescope, having previously been UK project manager for its photometric calibration. He remembers that he *'Did not know nearly enough about the project, but over the next three weeks or so he had several meetings with Gordon Carpenter both at ROE and at his home in Craiglockhart'*. Carpenter explained the flux collector concept, the contractual arrangements already in place with Dunford Hadfields and Grubb Parsons and outlined the plans for the building and dome in Hawaii. Colin's main interest was in the optical work and the performance of the telescope so, since it was clear that Carpenter wanted to continue developing the various arrangements that he had already started in Hawaii, it was agreed that Humphries would be responsible for the UK side of things, notably the optics and the telescope structure.

So, over the next few months, Colin made visits to Grubb Parsons and to Dunford Hadfields to familiarise himself with the work that they were doing and to get to know the project staff at both Newcastle and at Sheffield. He recalls that it was during these visits that he became interested in the optical specification of the primary mirror and the performance of the optical support system. He wanted to know at this very early stage whether the primary mirror would in principle be



J.K. Davies, *The Life Story of an Infrared Telescope*, Springer Praxis Books,
DOI 10.1007/978-3-319-23579-0_3

capable of producing diffraction limited images at visible wavelengths rather than just in the mid-infrared as originally intended. During the winter Colin was in contact with David Brown at Grubb Parsons and John Pope at the Royal Greenwich Observatory (RGO). Pope was able to assist in the analysis of the support system since the RGO had just obtained a CAD system that could calculate structural deflections and confirm the original calculations done by Imperial College. This was vital since the mirror cell performance was just as important as the mirror itself if the best possible image quality was to be achieved.

Others on the team included software engineer Bryan Bell and Ian Sheffield. Ian had joined ROE in the late 1960s and had been working in the electronics lab as a general electronics technician when he was transferred to the flux collector project. He and Gordon Adam worked on the monitor screens that would display the status of the telescope and indicate where it was pointing. Those screens used cathode ray tubes and, at the time, there were no commercially available units guaranteed to work in the thin air at 14,000 ft. A local company produced some specialised units which were tested in an altitude chamber at the Ferranti company's environmental test facility at Crewe Toll in Edinburgh. It was not only the potential problem of breakdown of the high voltage that concerned them; there was also the issue of cooling. Most electronic components require cooling which is usually done by convection; there was a worry that in the rarefied air at the summit the electronics might overheat. The special units seemed to work quite well, but there were some breakdowns once the equipment arrived in Hawaii. In the end the project reverted to using commercially available equipment, which seemed to work quite happily at the summit.

In 1977 mechanical technician Alf Neild joined ROE from the SRC laboratory at Daresbury where he had been working for 7 years on nuclear physics experiments. According to him *'The funny thing was, it was my Daresbury boss who insisted that I applied for the Telescope Technician job, having never seen a telescope in my life before. But I got an interview and was offered the job in April 77'*. He spent the next 8 months living in bed-and-breakfast accommodations; commuting between Edinburgh, Sheffield and Newcastle upon Tyne.

## Building a Building

A telescope requires both a building and a dome and in January 1975 the project was hoping to start pouring concrete for the building on Mauna Kea later that year, but it was awaiting the approval of the environmental impact statement. The building design started off as rectangular with the dome offset to the east, but it evolved into a more circular shape with the curved wall of the dome on the western edge and the mirror handling area and support building squaring off the north and east sides. The new design, according to Carpenter, *'looked better architecturally'* in that it echoed the other dome-like buildings along the summit ridge. The precise shape of the building footprint also took in the need for the provision of a road to the site of the NASA IRTF, which was to fork off just before the building and cut along below the site. Four possible internal layouts were proposed and thermal control loomed large in the discussions, with a consensus that observers ***might*** require some heating for comfort and this problem would have to be solved by ducting the resulting warm air away. A viewing gallery for visitors and a catwalk were included, but those would soon be removed as funds began to run short.

A contract for the design of the building was signed with Alfred A. Yee and Associates of Honolulu, but by then the environmental lobby had caused plans for the mid-level facilities to be scaled back so there were knock-on concerns as to what accommodation would be needed at the summit. It seemed likely for a time that the building would have to contain facilities for observers eating and sleeping, but it was clear that workshop personnel could not be expected to live at the summit. Nonetheless, in a letter to Gordon Carpenter on the 30th of December 1974, Roger Griffin said that he was pleased that the building plan included a bedroom since he believed that some observers would prefer to stay at the summit even if they have to '*Heat their own soup instead of being waited on*'. The same letter mentions a spiral stair that could be extended to reach the flat roof and indeed according to Richard Jameson '*For quite a long time the plan was to have a slide-off-roof building, Jim Ring was in favour but he was outvoted so we finished up with a dome, but a very small dome*'.

**Fig. 3.1** The building foundations in June 1976 (Photo Ray Wolstencroft)

In early 1976 Ian Robson, an astronomer at Queen Mary College, London who had joined the steering committee in 1975, visited Hawaii to report on progress and to consider options for downsizing as money was becoming rather tight. Ian made a number of recommendations, including that the proposed lift could be abandoned and, if necessary, the upper and lower doors could be used for moving any equipment between the observing and basement floors.

**Fig. 3.2** Vincent Reddish at the telescope construction site (From the ROE archive, undated but probably 1976)

Several British firms had been contacted about constructing the dome which would cap the building but no suitable expertise existed in the UK and so the project evaluated three potential American suppliers; the L & F Machine Company, Observa-Dome Inc and Astrotec. After a budget increase of £400,000 had been obtained to provide a dome of 58 ft (17.4 m) clear Internal diameter and a building to match, a contract was placed with Observa-Dome Inc of Jackson Mississippi towards the middle of 1975. Progress was such that both the building and the aluminium panelled dome were substantially complete, although not yet joined together, by the autumn of 1976.

**Fig. 3.3** A September 1976 view of the construction site with the building walls and floor starting

## Instrumentation Plans

Since a telescope without instruments cannot do astronomy, in parallel to the telescope construction an instrumentation programme was being carried out. Terry Lee, who would eventually play a key role in the project, had no formal involvement at this point but he met with Carpenter from time to time, to act as his *"tame astronomer"*. Terry had come to Edinburgh in 1972 from the Culham Laboratory, where he had been part of the Low Temperature Astrophysics Project and had joined the ROE staff as a researcher on interstellar gas-grain processes. He then became involved with data reduction for an infrared interferometer, which had been bought by Michael Smyth of the University of Edinburgh. When it was decided to fund an infrared instrumentation programme at ROE, Terry was made responsible for this group. He was joined by David Beattie, optical engineer John Harris, a software engineer called Malcolm Stewart, astronomer Peredur Williams, Mike Allen, an electronics specialist who had recently returned from working in Australia and another electronics person, Bill Parker. One other member of the group was Gordon Cork, who was working on a digital electronics system for transforming the output of the Michelson interferometer system to spectra for Michael Smyth and Terry, but who died suddenly in 1976. It was a time when infrared instrumentation was changing rapidly, notably with the development of Indium-Antimonide (InSb) solid state detectors for the near infrared (1–5 μm) region which were much more sensitive and had lower noise than the lead sulphide detectors then in use. It was vital to understand how these developments could be exploited for astronomy, so a detector test programme was put in place at Imperial College.

Not just infrared instruments were under consideration, as a flux collector with a relatively low quality mirror it was expected that the telescope would be used for sub-millimetre observing. However, when Ian Robson joined the Steering Committee one of the first things he did was to cancel the folded flat secondary mirror that was intended to feed the sub-millimetre instruments. These instruments had originally been intended to sit above the primary mirror and be fed by the fast uncorrected primary beam of the telescope. However, Ian realised that it would be so difficult to mount the instruments there, and to gain access to them for liquid helium filling, that they would be very unpopular and hence would be allocated very little telescope time irrespective of how good a scientific case could be made for them. He suggested to Gordon Carpenter that it would be possible to make the sub-millimetre instruments fit at the normal Cassegrain focus and Gordon was delighted to be able to remove this extra engineering complication. However, the 1 m diameter flat mirror had already been purchased and to this day Ian suspects that someone probably got a very good coffee table out of it.

Another key decision, based on advice from people who had observed on telescopes in the United States, was to develop "common-user" instruments that did not require specialist knowledge to operate. This was an essential step to involve traditional optical astronomers in the new and rapidly developing field of infrared astronomy. Along with these common-user instruments would come computer systems to control them and to log and reduce the data in real-time. These systems would replace the traditional strip-charts, in which a pen traced detector output onto paper, with digital records that could be processed by computers and assessed during the night. This would reduce the mental load on observers who had to deal with both the normal fatigue induced by jet-lag and working at night, compounded by the as yet unknown effects of altitude.

As part of this process, a telescope simulation facility was developed at the ROE. The simulator enabled instrument builders to bring their newly developed instruments to Edinburgh and try them out. Terry Lee felt that '*A lot of instrument builders thought that if something did not work in the labs then if they took it to the telescope it* ***would*** *work. A lot of telescope time was wasted in this way and we could not afford to do that on Mauna Kea. So we built a rig to tilt photometers and had an optical setup to direct an optical or infrared beam into the instrument under test'*. The first simulator was built in the ROE Villas building and could provide f/9 or f/35 beams of optical or infrared light to feed an instrument while it was tilted to test for any internal flexure of its components. The original simulator no longer exists, but since then progressively larger ones have been built and remain in use in Edinburgh to this day.

The instrument suite planned for the telescope was to include the following

- Photometers using InSb detectors with multiple apertures, filters and Circular Variable Filters (CVF) covering 1–5.6 μm.
- A cryostat with a bolometer detector and filters covering 3.8–40 μm.
- A cryostat with a bolometer or doped Si detector, filters covering 8–40 μm and a CVF covering 8–14 μm.
- A polarimeter insert to be developed in collaboration with what was then Hatfield Polytechnic.
- A fast photometry capability.
- A visible photometer covering the standard UBVRI bands.
- An infrared cooled grating spectrometer.
- High resolution Fourier Transform Spectrometers (both optical and infrared instruments would be attempted, only the optical version would ever reach the telescope).

There were a number of crucial decisions to be made including the configuration of the instruments. Infrared detectors must be cooled to their optimum working temperatures, i.e. to liquid nitrogen temperatures for operation in the near infrared and to below the natural boiling point of liquid helium for longer wavelengths. Normal practice was to build the cryostat so that the working surface was the bottom face of a cryogen tank, an arrangement that is optimal for cryogenic performance. This layout also works well for mounting the detector and optical components with the light beam entering through the side of the cryostat. However this arrangement means that the light from the telescope must be bent through 90° to take it out sideways into the cryostat. This requires an extra warm mirror which is not ideal because it emits in the thermal infrared and degrades the performance of the telescope/instrument combination, a problem that gets worse if the mirror is dirty and so is more emissive. However, upward looking cryostats were more complicated to build and had relatively short hold-times. Furthermore, they needed to be dismounted from the telescope to be filled, a time consuming and potentially risky operation.

The side-looking configuration had the additional advantage that if the tertiary mirror is dichroic (i.e. it reflects infrared wavelengths while transmitting in the visible) then the visible light passing through it can be used to see where the telescope is pointing. So a TV camera, mounted on a motorised X-Y slide (called a crosshead), was installed at the bottom of the instrument package. With this it became possible both to find objects of interest and to do offset guiding, using a star near to the, possibly invisible, infrared target to maintain the telescope's pointing as it tracked across the sky. This was an important consideration for a telescope of new design whose acquisition, tracking and offsetting properties were not predictable. Furthermore, if this dichroic mirror is mounted on a hollow bearing concentric with the telescope axis, then several instruments can be mounted on the telescope at any one time, providing backup in the case of instrument failure and offering a choice of instrument based on observing conditions and scientific priorities. The longer hold-times of the sideways looking cryostats also allowed refilling operations to become a routine daytime task and not waste time during the night.

For their instruments the ROE bought 6 cryostats from Oxford Instruments to a design that had been developed in conjunction with them. This design featured long hold-times, flats on the outer case for mounting motors, windows and fittings for pumping on the inner vessel. The InSb detectors were obtained from the USA, either from the Santa Barbara Research Corporation or Cincinnati Electronics. Interference filters were purchased from OCLI Europe, who at the time had a base in Fife just across the Firth of Forth from Edinburgh. Rather than hunting for the closest available existing filters from surplus catalogues, the team specified filters to match the atmospheric windows and ordered 50 sets of them. They then sold, or even gave away, about half of the sets to IR groups around the world. This bulk purchase both saved the project money and crucially meant that a large part of the infrared astronomy community was using the same set of filters. This was a decision that had profound implications for infrared astronomy around the world.

One of the instrument builders was David (Dave) Robertson, who had applied for a job as an electronics engineer at the Observatory in 1977, admitting that one of the things that attracted him was the possibility of travel to Hawaii. He started on the 1st of July 1977 and remembers that there were two groups who were working on the project, which he describes as *'UK-IRT, which was the telescope group and UK-IRI, which was Terry Lee's instrumentation empire and which seemed to run quite independently of anything else'*. So Dave got involved with Terry Lee and the rest of the UK-IRI group.

Other instrument options were also considered. On 6 April 1978 Terry Lee circulated a letter seeking proposals for instruments to be set up in the telescope coudé room. A Michelson interferometer, an FP interferometer and an infrared Michelson interferometer were all mentioned as possibilities, but not everything could be accommodated. On the 12th of April Terry wrote to Roger Griffin, who had proposed a radial velocity instrument in a cylinder about 6 feet long to say that space was already in great demand and there might not be enough room for a '*coffin sized instrument*'.

## Building the Telescope

Construction of the telescope structure proceeded at the Special Steels Division of Dunford Hadfields in Sheffield and some of the telescope team would spend much of the next year shuttling between there and Edinburgh, going down for a week or so whenever a series of tests were scheduled. Ian Sheffield found himself commuting up and down regularly. His task was to assist in the construction of the telescope and to be the main ROE point of contact for the electronics. Ian was expecting to be the main electronic technician for the telescope in Hawaii so it was both a familiarisation process for him and a chance to check that what the contractors were producing seemed reasonable.

**Fig. 3.4** Welding the telescope yoke (Photo from ROE archives but origin unknown)

Bryan Bell produced the low level subroutines for telescope control and the linkages in Basic. He had been recruited by ROE to develop the telescope hardware and software, but worked exclusively in Sheffield before moving directly to Hawaii. While Bryan was working on the detailed software Ian, who was interested in programming, put together menu driven test programs that came up on a screen and made it possible to enter astronomical co-ordinates (Right Ascension and Declination) or telescope co-ordinates (altitude and azimuth) and drive the telescope around. According to Colin Humphries, Ian's contribution '*was important because, without it, we would not have been able to achieve what was needed for testing the performance of the telescope in the short time available before it went over to Hawaii*'. Indeed Colin says he was so impressed by what Ian had done that he queried why it was necessary for the final software suite to take so long to develop when Ian's software had been produced in a very short time.

Terry Purkins, who joined the team in 1977, described the Sheffield factory as '*Very broken down, just like that in the opening scenes of the film The Full Monty*' and indeed most of the site was used for large-scale industrial processes, which created a number of challenges. To try and keep the iron oxide dust, which permeated everywhere, off the computers and other electronics they were installed in a big aluminium trailer, probably an ex-military vehicle, with dust seals on the door. However, it never really worked; whenever the door was opened there was red dust everywhere. Even electricity was a problem; the computers were designed to work at 60 Hz because they were going to America, so a frequency converter from 50 to 60 Hz was needed. In those days this was not done with electronics; instead a motor-driven generator with a control on the side, like a hand throttle, was used to alter the speed. Three-phase mains power was fed in at 50 Hz, which drove the generator motor and the speed was altered with the hand throttle until something approaching 60 Hz was achieved. This was then used to drive the telescope and electronic systems under test.

There were two other, even less obvious hazards. The telescope was parked in a big hanger, of which only about 3/4 was closed off. The rest of the hanger was used as a dumping ground for the refractory glass which was used to line the steel foundries. Ian Sheffield remarks that '*There is nothing so mind concentrating as sitting at a computer terminal when suddenly behind you 2 or 3 tonnes of glass are dumped out of a tipper lorry with a big crash*'. Another unusual risk came from a railway siding along which wagons of molten pig iron were parked to cool down. Visitors were told never to park a car anywhere near them because the heat would take the paint off the side of a car facing the trucks.

Preliminary static and dynamic tests on the structure were carried out in May of 1977 by the ROE project team assisted by some consultants. The requirement for further mechanical and electrical work on ancillary parts and the control systems caused revisions to the delivery schedule and a postponement of the shipping date from mid-July to mid-December, but by the early autumn the telescope and its control system were substantially complete.

On the 21st of June 1977 there was a press conference in the factory at which the completed telescope structure was shown off. It was at that time, and would remain for many years, the World's biggest infrared telescope, and it was British designed and British built. The telescope and its control system, fitted with a dummy concrete mirror weighing the same as the real thing, were completed in Sheffield by September 1977. This allowed the mirror's axial support system to be tested using the 80 pad pneumatic control system, which supported it in its operating condition, floating 5 mm above its rest position. Four axial load-cells under the mirror sensed and corrected for any tilt of the mirror. The testing was performed by Adrian Taylor of the Department of Mechanical Engineering at Sheffield University. Radial positioning of the primary mirror was provided by 24 counterbalanced lever arms in contact with the Invar pads attached to the outer edge of the mirror and pivoted at points attached to the mirror cell. These counterbalanced arms automatically compensated for gravitational effects when the telescope was tilted away from the vertical. By producing a push on the lower side of the mirror and a pull on the opposite upper side of the mirror the arms would keep the mirror centred radially for all pointing directions of the telescope.

**Fig. 3.5** The telescope structure assembled in the factory in Sheffield. Gordon Carpenter had offered several options for the colour scheme and Ian Robson had been very supportive of the modern orange and blue look, rather than something more conventional like a drab grey or green (Photo Colin Humphries, who pasted in an image of himself at a desk to give an idea of scale)

A prolonged period of acceptance testing was then carried out. The tests included measurements of the flexure of the main components, frequency responses in both motion axes, calibration of the 20-bit digital encoders that determined the telescope pointing and tests of the drive and servo control systems. The objective was to be as sure as possible that the telescope would meet its specifications, bearing in mind that no optics were available at that time. So the structure was put through a series of alignment, deflection and movement tests, with the telescope driven electrically under partial computer control. However, without any real optics, it was only possible to test to the limits that could be verified mechanically using a laser to measure flexure. The telescope, still loaded with its concrete mirror, was driven to all positions and the flexure of the system was measured with the laser.

## Improving the Mirror

By about October 1976, although polishing of the secondary mirror was at an early stage, the optical figuring of the primary had reached the point at which Grubb Parsons were confident that the mirror would eventually give the optical performance specified in the contract. They also indicated that further polishing work would be required if the possibility of upgrading the mirror specification was to be taken forward. So, on the 28th of October 1976, it was reported to the IRFC Steering Committee that the primary mirror could be polished to produce 1 arcsecond images (effectively as good as a normal optical telescope) for comparatively little extra cost. Richard Jameson, who was on the committee, remembers, *'Out of the blue Grubb Parsons said they could do a 1 arcsecond beam for not much more money, which with a few small sacrifices we had, so we accepted. There was a brief discussion as to whether this was a good idea due to the danger that optical astronomers might try and highjack our infrared telescope'*. However, it clearly was a good idea (although the people involved may not have realised just how good an idea it would eventually turn out to be) and with the work by Colin Humphries on the mirror specification and by John Pope confirming that the support system was good enough to merit such an improvement, the steering committee asked the ROE director to request the additional funds. So, on the 1st of February 1977 a memo from the Director of the ROE to the ASTRONOMY II committee asked that the budget be increased by £12,000 in order to secure a more accurately figured mirror. At about the same time Colin was formally made deputy project manager for the telescope.

While the extra money was being sought, the polishing work progressed and the original "flux collector" specification was close to being met by March of 1977. Colin Humphries made a visit to Grubb Parsons, saw the test results showing that the mirror would be diffraction limited at about 8 μm and agreed that the contracted specification had been met. However, before he left he asked David Brown, the Technical Director of Grubb Parsons, whether he felt that additional figuring of the primary mirror surface could indeed be done to improve the optical performance even further. Terry Purkins remembers that one such conversation took place while '*having dinner in Newcastle one evening with Colin and David early in 1977. We had reached a stage where David Brown said he was very confident that Grubb Parsons could greatly improve the figuring of the primary, well beyond the agreed specification, if we agreed to pay £12 k for further polishing. David made it clear he wasn't guaranteeing anything other than exceeding the specification, but felt the risk to us was worth taking since we would effectively be improving a flux collector to the status of an optical telescope – with far reaching consequences for the distant future. That was the night I personally was convinced we should take the risk but it was Colin who had the financial authority and to his credit he did push hard for the improved specification and willingly took the risk*'. Brown soon confirmed in a letter that, in his opinion and for an agreed cost to be borne by SRC, the performance could probably be improved. He estimated that the cost of making this attempt would be around £10,000.

During April and May 1977 the details of this proposal were worked out between Brown and Humphries, specifying the criteria for measuring the improvement in the mirror and developing a formula for calculating the cash value of any improvement achieved. Although encircled energy diameters had been used up to that point for specification purposes, the specification for the additional optical figuring work was expressed entirely in terms of a modulation transfer function (MTF) at specified spatial frequencies. A MTF description was considered more appropriate as the performance of the mirror was improved. It also allowed the results of shearing interferometric testing to be interpreted more meaningfully rather than using parameters based only on geometric considerations. Additionally, the performance could be compared with existing MTF measurements of the atmospheric seeing conditions at the Mauna Kea site over the wide range of infrared wavelengths at which the telescope would be used. In terms of an encircled energy diameter the improved specification based on MTFs was equivalent to obtaining 95 % of the incident energy from the primary mirror within a diameter of 1 arcsecond.

In a letter to David Brown on the 3rd of May 1977 Colin Humphries accepted that the risk of this extra polishing would fall on the ROE and affirmed that the minimum guaranteed performance would remain the basic specification already agreed with the SRC. A contract amendment was proposed by Grubb Parsons on the 17th of May 1977 and accepted not long afterwards. The precise chronology of the decision to proceed further and its ultimate approval are hard to reconstruct but according to Colin '*By September 1977 the Grubb Parsons contract for the optical work on the primary mirror had been completed. At this time I then made the proposal to improve the mirror figure even further if it was considered possible. I accepted David Brown's quotation for this work which was started immediately and was finished in October 1977. Grubb Parsons had other contracts waiting to be started.. [which is why].. it was essential to take an immediate decision to go ahead with the improved figuring, otherwise Grubb Parsons could not have incorporated it into their work schedule*'. The additional figuring to give the improved specification was achieved in just two figuring cycles and polishing of the mirror was finished by the autumn of 1977. The agreed cost increment for the improved image quality was set at £9325.

**Fig. 3.6** Two images of the mirror being polished (Photo from ROE archives, but origin unknown)

According to Colin *'The decision to go ahead with further figuring of the mirror was not without risk. With such a thin mirror this was unknown territory and there was a possibility that it could result in the slope errors of the surface already achieved increasing rather than decreasing. However, David Brown at Grubb Parsons was one of the foremost optical engineers in the world and his knowledge and experience of producing large, high quality, astronomical mirrors was second to none. It was therefore considered safe to rely on his judgement'*.

By the time of the next meeting of the Steering Committee early the following year the primary mirror and much of the project team was already in Hawaii, but Colin returned to London to present a paper bringing the committee up to date on progress with the telescope and the mirror. He also asked for retrospective approval of the decision to improve the mirror figure which, considering the small cost, he admits that he '*rather took for granted*'.

## Tragedy Strikes

On the 1st of June 1977, in the middle of the mirror polishing process, the project suffered a sudden setback when Gordon Carpenter had a heart attack at his desk in the ROE management building. Alex McLachlan, Ian Sheffield and site first aider Stan Blackley all attempted to revive him but it was too late, by the time they found him Carpenter was already dead. Colin Humphries was promoted to oversee construction of the telescope and was pitched into the hot seat with a lot to learn in a very short time. Colin puts it thus *'It was only a few months after my first involvement in the UKIRT project that on a sunny day in the summer of 1977, when I was at home, I received a telephone call from Vincent Reddish informing me that Gordon Carpenter had died and asking me to come immediately to ROE. At that time I was ill-prepared to take over the project management because I had been involved for only a few months and, although I was gradually becoming familiar with the work in the UK on the mirror development and the telescope structure, I knew very little about the Hawaiian end of things'*.

Colin added Terry Purkins to his team because the two had worked together previously. Terry had joined ROE in 1967 and worked on S47, an experiment to carry out ultraviolet photometric studies from an instrument carried to high altitude by a Skylark rocket launched from Sardinia. He then worked as a member of a team of four on the S2/68 space experiment for the European TD-1A satellite for a few years until that finished in 1972. He had been involved in some early testing of the 60 inch IRFC in Tenerife and in the latter stages of the installation of a ROE-built 24 inch Cassegrain telescope in Monte Porzio, near Frascati in Italy. So although Terry had never sought to work in the telescope construction team, he became involved anyway and soon found himself travelling to Sheffield most weeks for the rest of that year.

## Putting It All Together

Once the optical work on the primary mirror had been completed the realisation came that, under the existing plan, the primary mirror from Grubb Parsons and the mirror cell from Dunford Hadfields would not come together until they arrived in Hawaii. Since the axial support system constructed by Grubb Parsons and the mirror support cell built by Dunford Hadfields were not identical, for example the telescope mirror cell had radial as well as axial supports for the mirror whereas the optical testing system had axial supports only, it seemed to Colin Humphries that this presented an unacceptable risk. If there were to be any problems in the combined performance of the primary mirror and its support cell when they were assembled, it would be better to check this before the components left the UK rather than discover them once everything had arrived in Hawaii. So Colin contacted Des Hickenson at Dunford Hadfields to point out the wisdom of combining the mirror with the actual mirror cell before sending them to Hawaii. He asked him to arrange for the mirror support cell to be transported to Newcastle so that the combined system could be tested optically in the test tower at Grubb Parsons. There were some commercial confidentiality design considerations on the part of Dunford Hadfields, but Hickenson agreed and the mirror cell duly arrived in Newcastle where the combined system was tested in November 1977. Fortunately, no problems were detected and it was possible to do some additional optical testing, for example, the amount of astigmatism that could be induced by an imbalance in the lateral forces provided by the 24 radial levers around the outer edge of the mirror was measured. Roderick Willstrop, from the Institute of Astronomy in Cambridge, contributed to this part of the work as well as to the later testing in Hawaii. The works testing went on until the 11th of December 1977 to meet the planned shipping date at the end of 1977. Formal handover of the telescope to SRC was due in the first half of 1978.

## UKIRT Gets Its Name

During its early life the project, and the telescope, was generally known as the 3.8 m IR Flux Collector but it would eventually become known as UKIRT. Its is not clear if this name was ever formally chosen, but Alex McLachlan writes *'One day I received a phone call from the contracts office at Swindon to say that, in order to facilitate entry of the telescopes and related material into Hawaii, the US customs would require us to give them a code that we would use to identify such material. I suggested UKIRT'.* About 2 years later, when the details of the ceremony to inaugurate the telescope were being worked out, a proposal to name the telescope after the late Gordon Carpenter was floated, as was the name of James Cook, the British mariner who had played such a part in Hawaii's history. Although the idea of the "Carpenter Telescope" gained some traction for a while, neither idea was adopted, so UKIRT it remained.

# Chapter 4
# To Hawaii

## Preparing the Ground

As the telescope began to approach completion in the UK and the building on Mauna Kea was being finished, it was time to start planning the movement of both men and machines to Hawaii. An important financial aspect of this would be the tax implications of importing the equipment into Hawaii, since US customs duty was a potentially large cost. The Canada-France-Hawaii Telescope (CFHT) team had avoided paying duty by having a special act of congress (Public Law 93-630, 3 January 1975) passed to allow their telescope and its supporting hardware free entry in the USA. In London the SRC administrators looked into both this and other options and did indeed find a solution. Accordingly on the 18th of March 1977 US customs authorised import of UKIRT components free of tax noting that these components would remain the property of the British Government while in the United States.



J.K. Davies, *The Life Story of an Infrared Telescope*, Springer Praxis Books,
DOI 10.1007/978-3-319-23579-0_4

**Fig. 4.1** The UKIRT dome under construction in April 1977 (Photo Ian Robson)

The plan had been for Gordon Carpenter and Alex McLachlan to go out to Hawaii to help set up a local office and prepare the ground for the arrival of the staff who would assemble the telescope. Despite the suddenness of Carpenter's death in June 1977, McLachlan was on his way just a fortnight later. Looking back Alex says *'It hardly seems possible, but thinking about it now Gordon Carpenter was an excellent manager, really one of the best. He had already had a heart attack and a minor stroke between 1974 and 1977 but because he always kept the staff fully informed of what the plans were and what was going on we were able to continue with the project when he died suddenly. He'd been off ill a few times in that period, but because of his approach to management we were able to continue'*.

Alex and his family duly arrived in Hilo, the largest town on the Big Island of Hawaii, and were on their own for 6 months or so. They stayed for a short while in the Naniloa Surf Hotel before finding a house to rent and getting their two young daughters into a local school. The initial project office was in Suite 206, 100 Pauahi St which had recently been vacated by Fred Zobrist, the lawyer who had been acting for ROE. In the same building was Peter Sydserff the site engineer/site manager for the CFHT.

Alex McLachlan then spent a lot of time looking for a building to rent as a temporary sea-level base. Hilo is a close community and of course word of his search soon got around and he settled on a warehouse, which had a suite of offices attached at the front, at 900 Leilani St, close to the airport. Leilani St was in a nondescript semi-industrial area of Hilo with its main claim to fame being that at its end was the town's main rubbish dump, so for years the UKIRT headquarters would be described as being "The last building on the way to the dump". The office was leased for a period of 3 years with options to extend the lease if required. Alex was followed to Hawaii by Colin Humphries who had been appointed to be the first Officer-In-Charge of the telescope during the summer of 1977 and who arrived in October 1977. They were soon joined by Alf Neild and his wife who arrived in Hilo on the 4th of January 1978 and by David Brown, an administrative officer who, like Alf Neild, had come from the Daresbury Laboratory.

The choice of Officer-In-Charge in Hawaii had been made by an appointment board that included Vincent Reddish and Jim Ring. After the post had been advertised and interviews held, the board decided that instead of the usual 3 year tour of duty for such an overseas appointment, it would be split into two periods of 18 months each. Thus, Colin Humphries became the Officer-In-Charge until 31 December 1978, and Terry Lee from 1st January 1979. In this way it was hoped that the two parts of the UKIRT project, telescope construction and telescope instrumentation, would be able to proceed smoothly and contiguously on the shortest possible timescale. This arrangement did, however, put pressure on the construction project and Colin was not entirely happy because *'it meant that now there was a fixed deadline without contingency for the completion of the telescope construction; the first six months of the appointment would be spent mainly in the UK completing the large amount of work that was left to be done there'* and *'there would then be less than one year in Hawaii to complete the crucial on-site construction as well as the engineering and optical testing of the telescope'*.

Not all came willingly. Terry Purkins was asked to go to Hawaii in December 1977 to act as Deputy Officer-In-Charge during 1978. Terry refused to go on the basis that Colin Humphries had been temporarily promoted one grade on his appointment, but that he was not being offered a comparable promotion. Terry was called in to see ROE Director Vincent Reddish on the 23rd of December and recalls that *'It was made very clear to me that I was expected to go'*. Terry decided to play hardball and absolutely refused, saying *'You can fire me if you want but I'm not going to go unless I get promoted'*. One week later, to his complete surprise, he received a letter congratulating him on his temporary promotion. He then had just 5 weeks to organise the departure of himself and his family to Hawaii. It was all done very quickly and they arrived in February 1978. Terry says that *'For the rest of the year it was just sheer hard work with hardly any time off at all'*. Purkins acted as Chief Engineer, looking after construction of the telescope, including supervising the 4 or 5 Dunford Hadfields staff from Sheffield who were present for much of 1978. Colin Humphries managed all of the scientific matters and the politics.

**Fig. 4.2** Astronomer Ian Gatley outside the Leilani St offices (Photo ROE Archive)

Once established in Leilani St the team continued to grow. Transferring from the instrumentation group in Edinburgh was Bill Parker, who had arrived at ROE in 1969 from Ferranti, where he had been working in the Bloodhound missile receiver laboratory. Bill's first job at ROE had been to work with Gordon Carpenter on the electronics for the twin telescopes then installed at ROE while, at the same time, he was managing the electronics laboratory in which Ian Sheffield had worked. The electronics was, he says, *'More like my kind of job'*. He also found himself fixing the photometers used to measure star brightnesses on photographic plates and repairing other instruments and telescopes when necessary. Next he worked on various projects for the telescope at the ROE outstation in Monte Porzio, Italy, from where he moved on to designing and installing the drive for the 1.5 m IRFC at Izana and then to designing the drives for the XY stage to go on UKIRT. Then in 1977, he was *'Asked out of the blue to go to Hawaii'*.

Also in Hawaii by then were Ron Beetles, Bryan Bell, Rory Urquhart (a mechanical technician recruited to fill the gap to be left when Ron Beetles returned to Edinburgh after the telescope commissioning was complete) and Ian Sheffield. Between them they set about developing workshop facilities and making business contacts with the local community. They were joined as group secretary by Yolanda Boyce, described by one UKIRT staff member as a '*Soothing presence, a robust Hawaiian woman who practically exuded serenity. She always had a motherly smile and a calming voice, and she was an absolutely top-drawer administrator. Everyone relied on her very heavily, and practically everyone adored her'*. Yolanda was later joined on the administration side by Anna Lucas, who would mostly be responsible for sorting out logistics for visiting observers, while John Clark and David Robertson joined the technical staff in the summer of 1978.

Ian Sheffield organised a local carpenter to fit benches to the workshop so the team had somewhere to work. At first there was no test equipment, but there was a Radio Shack electronics store across the road from where Ian bought the basic oscilloscopes and test equipment they needed. Ian became friendly with the owner of the store and from time to time did some moonlighting repairing computers for him. Although he did not know it, this extra-curricular activity was going to create difficulties later.

## Promoting the Project

As the project progressed the details of the telescope began to be publicised around the UK and in Europe. In 1977 a specially made UKIRT model (which survived and is now on view in the visitor centre at ROE), was displayed at the British Genius exhibition in Battersea Park. This was part of the celebration of the Queen's Silver Jubilee and Her Majesty visited the exhibition on the 27th of May.

On the 9th of December 1977 another Royal Astronomical Society specialist discussion on infrared astronomy was held in London. The UKIRT steering committee were responsible for setting the programme, which included talks by Colin Humphries on the telescope, Terry Lee on the instrumentation plan and Mike Selby on detectors and the capabilities of the telescope once it was operational. Michael Rowan-Robinson previewed the use of the telescope in the sub-millimetre region noting that it would be the most powerful instrument of its type until the advent of the proposed UK millimetre wave telescope (which would eventually become the James Clerk Maxwell Telescope, the JCMT). Specific Instrumentation talks were given by Richard Wade (speckle interferometry), Dave Adams (imaging extended objects using multiple masks), M. J. Smyth on high resolution spectroscopy with Fabry-Perot or Michelson interferometers plus J. C. D. Marsh and Ray Wolstencroft on polarimetry. Dave Aitken talked about the proposed InfraRed Astronomical Satellite (IRAS) which he thought would provide a basis for UKIRT work far into the 1980s and also on low-resolution spectroscopy. Scientific talks came from Peredur Williams and Gareth Wynn-Williams. A few days later, between the 12th and 15th of December 1977, there was an ESO conference in Geneva on Optical Telescopes of the Future. At this meeting papers by Brown and Humphries on

"Design and Performance of the U.K.I.R.T. Optical System" and by G. C. Carpenter et al. on "Structural Design Considerations for a Large Infrared and Optical Flux Collector" were presented and later published in the conference proceedings.

**Fig. 4.3** UKIRT model (Photo via Terry Lee)

## Transport to the Summit and Assembly

The first UKIRT components were shipped from Liverpool to Hawaii via the Panama Canal in December 1977 and arrived in Hawaii on the 12th of January. The yoke and mirror cell arrived separately on the 26th of the month. The unloading and transportation of the yoke was clearly going to be the biggest problem. Just getting if off the ship in Hilo was a bit of a worry because it needed a large crane on the dockside to lift it and there was concern that the total weight of the crane and the yoke might be more than the pier could support. In fact it all went well and the yoke was unloaded without difficulty and was taken to the Kuwaye Trucking Company's yard in Hilo. It was stored there while plans were made for taking it through the town.

**Fig. 4.4** Unloading the yoke at Hilo docks (Photo ROE Archive)

Moving the yoke through Hilo would be quite an enterprise in itself because it would completely occupy the width of the street and this operation attracted considerable local interest. Colin Humphries was very aware of local sensitivities regarding developments on the summit and wanted to be very careful about any press coverage, so he told his staff not to speak to journalists about it. Accordingly, when a reporter phoned up from Waimea asking about details of the movement of the telescope through the town, Alex McLachlan recalls that he replied *'I'm sorry I'm not allowed to speak to you about this'*. A few days later the paper came out with an article saying that Mr McLachlan of UKIRT, the staff member responsible for getting this though Hilo, is not worried that it might cause damage to downtown buildings. Alex had said no such thing, it seems that the journalist, unable to get an official comment, had simply made one up. The move was made on the morning of Sunday the 5th of February. The team met up before sunrise at the trucking yard and set off. The yoke was a tight fit at some places, but it was most impressive to see it go through the town.

**Fig. 4.5** The UKIRT telescope yoke on its low loader turns into Waianuenue Av in Hilo (Photo ROE Archive)

When the convoy reached the steep stretch of highway going from the Saddle Rd up to Hale Pohaku a big dumper truck was used to tow the low-loader and the yoke the rest of the way. As they approached the summit there were times when the yoke was sticking out over the edge of the road so it was quite an exciting time for all concerned. Other components followed over the next few weeks with, according to Colin Humphries '*Only a few instances where the removal of road signs or re-positioning of overhead power lines was required, mainly in the Hilo suburbs*'.

**Fig. 4.6** The telescope yoke on the dirt road to the summit (Photo ROE Archive)

On the 15th of February the low-loader lorry carrying the yoke, was backed carefully into the dome via its big roller doors. The weight of the yoke plus the lorry was of course far more than the dome floor would normally be expected to support so the floor was braced from underneath with steel poles. A week later a crane was positioned outside the dome and its hook lowered through the dome slit. The yoke was lifted by the crane and carefully positioned on its two piers. The primary mirror arrived in Hilo on the 10th of February and was moved to the summit on the 23rd. For Alex McLachlan the only drama during the setting up of the telescope was when someone fell off the dome and had to be rushed down to Hilo hospital. *'That was my first experience of the American system in that they wanted to know who would pay. I had to speak to the manager of the hospital and assure him that the bills would be paid by the British government'.*

This accident aside, the erection of the telescope itself went very smoothly. Colin Humphries recalls that *'The physically demanding task at high altitude of re-assembling the telescope structure was accomplished by a team of only four persons – Des Hickenson, Alex Gaymond and Les Wilson from Dunford Hadfields, together with support from Alf Neild of the UKIRT project group. This small team operated typically by spending periods of about 10 days at Mauna Kea followed by a two or three day break in Hilo at low altitude, and then repeating this routine. They had to organise the presence of mobile cranes at Mauna Kea when required. Later, Alan Deeley from Dunford Hadfields arrived to commission the telescope pointing control system that he had designed'.* Assembly of the telescope structure was completed in April 1978. The remaining period of that year was devoted to installing the optics and testing them under real observing conditions, and to refining the telescope control system so that the pointing performance of the telescope and its stability were optimised.

## A Dicey Few Moments

One day in about February 1978 Alex McLachlan, Alf Neild, Terry Purkins and Colin Humphries were working in the dome carrying out tests of the telescope when the weather started to deteriorate and it started to snow. The UKIRT team got word that the CFHT staff and others were getting off the mountain because the weather was bad, although in retrospect it is not clear if this referred to the snow or simply to the cloudy sky making astronomy impossible. Alex recalls urging the team to leave and after a while they reached a suitable break point in their work and agreed to set out back to Hale Pohaku. It was Alf's turn to drive down and although the snow had stopped by then, the road was icy with a light covering of snow. The truck was a three speed, automatic, seven seat Chevrolet Blazer, with an automatic choke. On start up, and due to the very low air temperature; the choke was fully open, making the engine revs over 2000 rpm. Due to the treacherous road conditions, Neild decided to select second gear, and see how fast it would take them. The four men got on board, and started the descent. The truck speed was safe until it passed the

"Drive With Care" sign; and entered the steepest section of road. Neild takes up the story

*'Going downhill, in second gear; with the choke still fully on, I quickly realised that the truck speed was too high. (About 10 mph). So I quickly changed down in to first, but the change down did not slow the vehicle as we were just sliding straight down a long left corner. The near side wheel struck something, which spun the truck counter clockwise; and we slowly crashed in to the bank'. (There was a 100 m drop, on the other side).* The vehicle finished up in the side of the hill facing the summit, the way they had come. The road was so icy that when one of them leapt from the truck he fell backwards onto the road. It was not possible to drive the truck up the incline, so with the help of Terry guiding him, Alf reversed the truck down to the first "switch-back", where he could turn the truck to point in the correct direction.

There is general agreement that it was a very dangerous incident. Terry Purkins was in the back of the vehicle and remembers that at one point he could see over the edge of the road and straight down the mountain side. He thinks that if they had gone over the edge they *'Would all have been severely injured and may well have died'*. Alf Neild feels *'That one incident could have killed half of the then UKIRT staff'* and Alex agrees saying 'I *was sitting in the back and I saw the edge of the road pass under the vehicle and I thought we were going over. I thought my time had come. It was a close run thing'*. Indeed one of the staff at the construction camp remembers that the occupants of the vehicle *'Came into the lounge after working late, looking very pale and said "We never expected to see you again" '*. This was a particularly vivid example of how dangerous the road to the summit could be.

Months earlier, there had been several incidents on the Mauna Kea road involving drivers of vehicles belonging to other telescopes. These had been a salutary warning to the UKIRT group who were about to have new staff, as well as members of universities and other establishments, who would be both tired and jet lagged on their arrival in Hilo. Therefore in February 1978, Colin Humphries and Terry Purkins introduced a rule that newly arrived staff or visitors would be allowed to drive the UKIRT vehicles only after they had experienced at least three return journeys to the summit of Mauna Kea as a passenger of drivers who were already approved. This rule remained in force for most of UKIRT's subsequent history.

Another consequence of the dangerous road was a decision soon after by Terry Lee and David Beattie to switch to using shorter wheelbase vehicles with manual transmission. Terry remembers that when the auditors discovered that the old vehicles had been disposed of before the regulation time or mileage had been reached he had been required to write a retrospective justification of this decision. However, even though the smaller vehicles were easier to handle and put fewer staff at risk, there were still to be incidents on the road from time to time.

## Power Headaches

Alex McLachlan still has his UKIRT hard hat, but it is something quite out of the ordinary. He explains *'The telescope wasn't properly balanced and this was a time associated with quite a lot of* [Earth] *tremors due to volcanic activity* [elsewhere on the island]. *We did not want the telescope to be stuck at an awkward angle when a tremor occurred, so we wanted to get the thing back to the vertical position as soon as possible once the tests were over. At the time we were entirely dependent on the power from the generators which were in a shed just across from the dome. I had spoken to the generator operator who was a University of Hawaii employee and said "Look whatever you do make sure we have power when we are carrying out these tests. Whatever you do please don't shut off the generator". So we were carrying out these tests and we had the telescope way over, top-end down and back end up when the power goes off. We couldn't believe it. I shot across to the generator shed. I'm sure my feet did not touch the ground and I really gave this man his character and got him to switch on the power again, which he did'.*

*Two weeks later we were working on the telescope and Terry Purkins wanted to clear some ice off the outside of the dome before we opened it up. He borrowed my hard hat and the hat blew off in the wind, went downhill into the snow and disappeared. The hard hat re-appeared a few weeks later on my desk in Hilo complete with this carriage bolt through it. It had been found by the person I had slagged off and he must have recognised the name and gave me a touch of his voodoo (laughs). I've had headaches ever since. Later I tried to cut through the bolt on the inside so I could wear the hat, but it proved impossible to do'.*

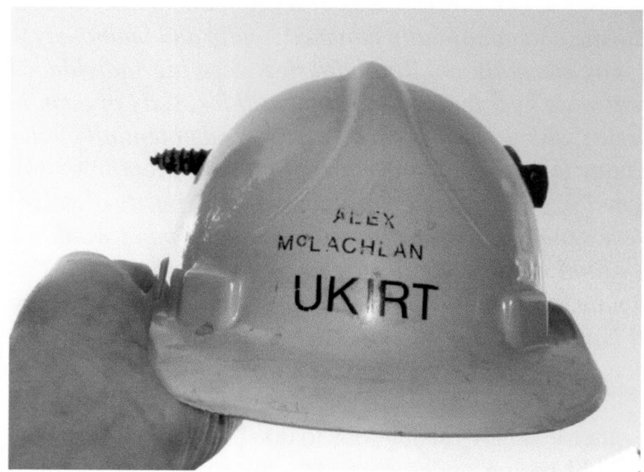


**Fig. 4.7** Alex McLachlan's hat (Photo Alex McLachlan)

## Outfitting the Dome

The electronics team of Bill Parker and Ian Sheffield did a number of diverse jobs, one of which might have saved Alex's helmet when they installed the power backup system in the basement of the telescope building. This comprised about forty 12 V batteries connected up to get 120 V at the end. They connected them all together, carefully leaving out links at random as they connected up the rest. They used an insulated spanner because when they got to the last link there was a potential of 120 V across it. Ian puts it in perspective *'I think the power was 25KVA so if you had dropped your spanner or screwdriver across the last two terminals when it was all connected it would have vapourised, along with you'*. They also installed a big 3-phase inverter in the basement. However, before it could be turned on it had to be connected up and, since it was a 3-phase supply, it was necessary to have a licensed state electrician do the final connection to the telescope. The contractor and Ian Sheffield had a wiring diagram and they studied this at sea-level and decided that *'the red wire goes there, this goes there and the rest of it'*. Then when they got to the summit they twice got it wrong. On the third attempt they almost got it right. When they were satisfied that it was set up correctly and charging the batteries, Bill Parker threw the switch and the telescope immediately stopped. Investigation showed that there was a changeover solenoid to switch from one system to the other and the coil of that solenoid was connected the wrong way around. The moment the mains went off the solenoid switched over to the mains and so did not connect the emergency batteries as it was supposed to. Ian describes it as '*Definitely a 14,000 ft problem, we went down to sea level and kept asking "why the hell did we connect it up that way?" It was purely due to altitude effects*'.

Bill Parker remembers the room below the dome where the inverter and battery were housed. He says '*The inverter was 50KVA, made by Elgar of San Diego and was three bays wide and continually hummed loudly and ominously with a grinding sound since it was eternally on. The batteries were big individual Lead-Calcium cells about 2 feet wide by 3 feet tall by 6 inches thick, sixty of them in two rows, all connected in series, about 132VDC in all and indeed potentially lethal.* There was a relay with its contacts exposed to the atmosphere and, according to Bill, *'Inevitably these were opened from time to time by very small deposits of volcanic grit. When the contacts were separated some brave soul (usually me, I wouldn't delegate it to someone else) would have to go down and clean the contacts with, maybe, a thin piece of non-conducting material while the inverter was still working, otherwise the inverter would have to be shut down while the contacts were cleaned. This was a hairy experience, trying not to come into contact with all the high voltage terminals while doing it'*.

There were other less hair-raising jobs to do. The control room had a raised floor so that cables could be run underneath. The floor was made from panels roughly 600 mm square and under each corner was a little support that held the floor up. The manufacturer said that, because of the altitude and the dryness of the summit, the corner of every tile had to be painted with tar to stop them from drying out. So one day a couple of people dutifully sat down and tarred every corner before the tiles were installed. Also, from time to time, the electronics staff would clamber inside the UKIRT yoke to do some wiring. They would get in at the north pier and walk, or rather stoop, around inside it to wire up the cables.

The biggest single problem that affected the telescope construction team concerned the dome and its crane. The dome control system was made by a company called MOOG, a name remembered well by Dave Robertson who always thought that they *'Made synthesisers for the music industry'* but of course it was a completely different company, MOOG hydraulic systems. The dome did rotate, but not very smoothly because it was underpowered and the sideways opening shutters often jammed. One day there was a power cut when the shutters were open so Bill Parker climbed up to the top of the shutters, where the motor was, and started screwing them closed by hand. There was no special handle, he just took a ratchet and put it on the end of the motor shaft and cranked. This was quite a feat in the thin air of the summit because he was fighting the friction in the unpowered motor as well as the inertia of the shutters. It took him an hour, and inevitably, just as he got them closed, the power came back on again!

Even more serious was that the dome crane turned out to be inadequate to lift the interchangeable telescope top-ends which carried the alternative secondary mirrors. The crane was originally specified to be able to lift 3 tons, but during the dome acceptance process on Mauna Kea no suitable test weight was available and the crane was accepted by one of the ROE project team without doing a test lift. When Colin Humphries and Terry Purkins became involved the crane was tested with a concrete block as a dummy load. During the test the lift cable jammed around some of its pulleys and one of them fell off, landing with a clang on the south column platform of the telescope. Closer inspection showed that a few of the cable strands had also broken. After this it was decided that no further tests would be carried out as the UKIRT staff considered that the crane could not be relied upon. Resolution of the crane problem would take until the summer of 1979.

## Storms

In about March 1978, with work on several telescope projects in full swing at the summit, there were various large and massive shipping containers lying around on the mountain when a powerful winter storm passed nearby. The site was evacuated for 2 or 3 days while the wind blew at a speed estimated to have been in excess of 125 miles an hour. Estimated is the key word here for when staff returned there was no sign of the weather station, it had been blown away. Probably about 3 miles away lay one of the shipping containers, with others scattered about nearer to the dome. The US Army brought in Chinook helicopters to recover the containers because they were inaccessible otherwise. Ian Sheffield saw one helicopter in action. *'The Chinook hooked onto one trailer while it hovered and then started to lift it. As the container began to lift it started to slide down the mountainside and you could see the pilot applying full power to stop the Chinook's rotor blades hitting the side of the crater'.*

## Aluminising

The telescope was initially assembled and tested with dummy mirrors. In parallel to this work the real primary and secondary mirrors were aluminised between the 28th of April and the 1st May 1978 using the coating plant of the nearby Canada-France-Hawaii Telescope. The mirror was transported on a flatbed truck from the UKIRT building to the CFHT dome, about 200 m away. The operation was carried out by a small team led by Colin Humphries and Terry Purkins with assistance from Bernard Cougrand who was responsible for the CFHT coating plant.

Since the chamber had never been used before, the aluminising of the UKIRT primary mirror also served as the commissioning and acceptance testing of the CFHT coating plant. So the group first had to do some "shakedown" tests on some large pieces of glass. It took two people over 6 hours just to load the chambers' vaporisation elements, not an easy task as it was necessary to stand on a platform and hook a set number of 1 cm long pure aluminium wires on the 100 or so tungsten vapourising elements. The tests were successful, and this made it possible to proceed with the aluminising of the primary mirror itself.

Alf Neild was there during this critically important process, as were some unexpected visitors. The primary mirror, still in its transportation box, was positioned on the trolley which would transport it in to the aluminising room. *'We had just lifted the steel lid off the mirror box, and were in the process of preparing to connect the crane to the mirror centre lifting ring; when we heard voices. Looking up, there was a squad of about ten American soldiers leaning over a barrier, looking down at us, from six floors up. Panic grabbed everyone, in case anything fell out of their pockets and dropped onto the mirror'.*

The whole process took some 20 hours, which included 12 hours to pump down the chamber to a near vacuum. In the event the thickness of the coating that was achieved was slightly thin (600–800 Å instead of 1000–1200 Å) but was adequate for what was required over the next year or so. The mirror was then transferred back from the CFHT to the UKIRT dome and installed in its mirror cell. By the second half of July the primary mirror and the f/9 secondary were safely installed and aligned.

## First Light

First light for UKIRT was at 03.23 on the 31st of July 1978. The target was the star Alpha Pegasi which appeared as a circle of light about 2 arcseconds in diameter. Getting the telescope onto target had taken a while because of errors in the encoders, which had not been set up correctly in the factory, but the images were good, and seeing them come out so well was something which Terry Purkins regards as the highlight of his time at UKIRT. The 4 inch finder telescope was installed on the 1st of August and testing of the telescope continued throughout the rest of 1978 at the hands of the ROE team assisted by R. V. Willstrop from Cambridge University. Checks of the alignment of the polar axis revealed that it had been set up within 15 arcminutes in azimuth and 1.5 arcminutes in elevation, errors which were quickly reduced by adjustments to the support columns.

Although he was not part of the optics group, for Ian Sheffield the most exciting thing was also seeing the basic telescope pointed at a bright star. For alignment they had a simple eyepiece and he looked though this and was just about blinded by starlight. He thought *'Yes, we've got a telescope now'*. However it was soon apparent that image quality was being degraded by both dome seeing resulting from the heat generated by all the activity in the building during the day, and an oscillation due to the characteristics of the encoders which produced an unwanted back and fro movement of up to 6 arcseconds in Right Ascension. This oscillation led one staff member to remark irreverently that UKIRT's pointing specification was to drift no more than 10 arcseconds per hour and that while it met that specification it also met a specification to drift no more than 10 arcseconds in half an hour and no more than 10 arcseconds in 10 minutes, and in 1 minute! There were also a number of other teething problems, such as flexure of the guide telescopes (which were eventually abandoned) and errors in the autoguider crosshead.

About the same time as the milestone of first light Terry Lee arrived in Hawaii as Astronomer-In-Charge designate. He was accompanied by David Beattie who would become his deputy in due course. Susan Beattie remembers that they were met and looked after by Cathy McLachlan who had already found them a house to rent, provided they took on the ginger cat which lived there. This mutual support was typical of the UKIRT operation for much of its life, with the established staff, and their families, chipping in to help new arrivals settle down. Terry's wife, Min Lee would soon find herself doing the same. This group identity of families, who did not have the common bond of working on the telescope itself, was perhaps partly a response to the many serious and long-lasting implications for a small expatriate community which needed to integrate itself into a local culture that was foreign in more than one sense of the word. Frances Hawarden, who would arrive a few years later believes that *'The UKIRT expatriate community was a pioneer of sorts in that it was among the first to be stationed overseas for scientific purposes – not for the usual diplomatic or commercial reasons'*. She believes that the contribution made to the success of the UKIRT presence in Hawaii by the various UKIRT employees' family members should not be overlooked. Susan Beattie would probably agree; within a short time of arriving she was asked by her husband to drive to the office, collect some large bits of metal and bring them to Hale Pohaku. Before long she was an ex-officio UKIRT driver, collecting VIPs from the airport and organising tours of the Hawaii Volcanoes National Park for visitors.

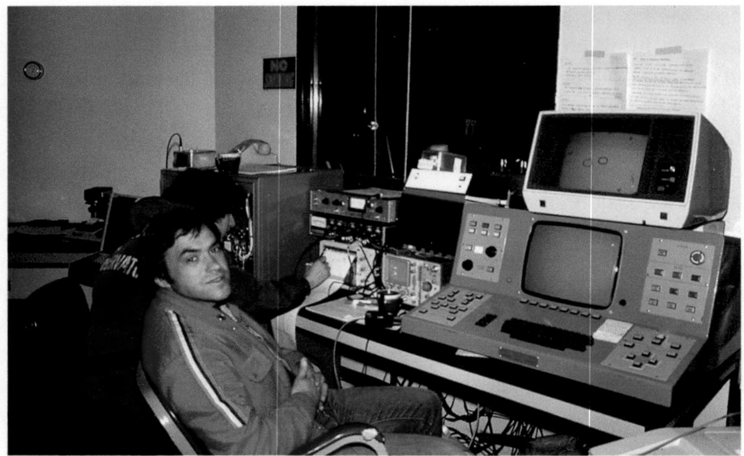

**Fig. 4.8** Terry Lee in the UKIRT control room. The photo is from March 1980, about 2 years after he first arrived in Hawaii (Photo Ian Robson)

Although Terry and David were not scheduled to take over until the end of the year, the two men had moved to Hawaii early because Colin Humphries had reported to the Steering Committee in May that the telescope assembly was going well and might be ready for visiting observers later in the year. Accordingly they felt that they had better get themselves and some instruments to Hawaii sooner rather than later in case they had to support visiting observers in the autumn. These arrangements had been agreed by Colin Humphries but he felt that it put him under further pressure to finish his part of the job on time. In the event David Beattie worked with Colin and Terry Purkins on acceptance testing while Terry Lee spent most of his time working on affairs at sea level. Indeed for a while Terry was not able to visit the summit, because one day he was 'dumped' heavily on the beach while body surfing. Colin drove to the hospital with Terry lying in the back of a station wagon where it was found that Terry had fractured two vertebrae, making trips up the bumpy summit road inadvisable.

## Hale Pohaku

Not just the telescope and its dome, but also the living arrangements at Mauna Kea's mid-level facility were fairly basic during the first few years of the UKIRT project. Early accommodation for the site testing teams had initially been provided in stone cabins (which gave the site its name of Hale Pohaku or "house of stone") but very soon afterwards a house trailer obtained from government surplus was installed in a fenced enclosure of the State Park Service. Then the company contracted to build the University of Hawaii's 88 inch telescope built a mess hall and a dormitory to accommodate their workers. These crude buildings were intended to serve for the duration of the construction phase (i.e., to 1969) and were expected to be replaced by a more sophisticated suite of buildings once the site began to develop.

Gordon Carpenter's 1974 report to the UK steering committee spoke of a new mid-level facility at the 8500 ft (2800 m) level to be built by about 1977 and which was expected to house about 70 people such as night assistants, daytime technical staff and the families of a few key personnel. His paper refers to both single accommodation for observers (costing $20 per day) and the rental of "houses". Similarly John Hutchinson's report of his visit to Hawaii in March of that year described IfA aspirations for shared facilities in an "observing village" to be built a little lower than Hale Pohaku, which would include houses, bedroom accommodation, recreation rooms, offices, workshops and laboratories. However, the approval of the CFHT, UKIRT and IRTF projects to be built on the summit began to worry some local people that the mountain was going to be overdeveloped and although the summit buildings were approved, plans for the further development of Hale Pohaku ran into problems, which caused the facility to be scaled back and delayed.

As a consequence of these delays, the UKIRT construction staff, and later the first generations of observers, had to make do with the somewhat spartan accommodation in the old construction camp for several years. This comprised a number of wooden huts, one belonging to UKIRT, which were reached by a short dirt road running off the highway just below where the new facility would eventually be built. The dormitory huts were fairly basic, not much above the standards of army barracks according to some, and were served by communal toilets and showers. There were a few small rooms, but the remainder of the early residents had bunk beds which could be curtained off into cubicles. Dave Robertson described these early accommodations as '*effectively a bunkhouse, there were no doors, just two beds on top of each other, open. One toilet, one shower room at the end, and that was for the day and night crews*'. There were no humidifiers and according to Alex McLachlan '*The first night I spent there the atmosphere was so dry that when I turned over the whole room lit up with sparks from the static electricity*'.

Another early resident, Terry Purkins, has a specific memory of these buildings. He spent a lot of time there when there was construction work going on and during which groups of contractors would be flown in from the main Hawaiian island of Oahu. According to him '*The air was full of cannabis smoke. The local contractors were all on pot all the time*'. In this Terry is supported by others, according to one regular HP resident 'The *atmosphere made it good for drying pakalolo, the Hawaiian word for marijuana/cannabis/pot. There were only four separate rooms in the CFHT dorm building, everything else was cubicles with bunk beds. One of the rooms was often out of commission because it would be full of large marijuana plants hanging upside down from the ceiling. (The workers liked to hang them upside down because that would, supposedly, make the resin in the plants seep down into the buds, which were the best smoking part of the plant. Concentrating the resin in the buds was the goal of a good pot grower.) The plant grew well in Hilo but would sometimes go mouldy if it was not properly dried, and Hale Pohaku was good for that'*.

Apart from the pot plants, there was another unusual feature of the early camp. When the UKIRT facility was first constructed, it had a (relatively) large women's toilet/bathroom and a small and cramped one for the men. There were far more men than women working on the mountain, indeed sometimes there were no women there for weeks on end, and after a few weeks the construction workers got sick of it, ripped the women's sign off the women's bathroom and posted it on the men's bathroom and vice versa. Thereafter, many women coming to the camp would ask why their bathroom had a urinal in it. Even with the signs, a combination of tiredness and altitude occasionally led to confusion. One of the few ladies to visit HP regularly in these early days remembers *'Coming down from the summit in the morning one time and being so tired that I went into the wrong bathroom. The doors were not marked with "men" or "women" but rather with the male and female symbols. I did of course realize that I was in the wrong place when I was in the bathroom stall and the person next to me had huge boots on, which belonged to a man, actually the camp manager at the time as I found out the next day'*.

**Fig. 4.9** A view of the huts at Hale Pohaku (Photo Harry Atkinson)

At first each organisation provided its own meals. The UKIRT team contracted with a company called Hawaii Island Safaris, which was run by JoAnn and Gordon Morse and was itself affiliated with another company called Holo-Holo Campers who, at least initially, brought up food from Hilo every day. Hawaii Island Safaris employed a cook called John (Johnny) Morris who, although widely believed to be British, was in fact an American with an English mother whose perceived Britishness came from him having lived in England for 14 years. Morris had done a degree in law in London, but had left England in 1973. He came to Hilo in 1974 and finished up working for Hawaii Island Safaris on Mauna Kea. According to him the company *'Had a mixed bag of workers. Some were good and some were marginal. Some were completely off the wall. We would stay out of the way of the crazy people and mind our own business until they flamed out and were sent down the hill'*. Morris had a better opinion of the UK contingent. According to him the team from Dunford Hadfields *'were super party people who on the weekends would enjoy the nightlife in Hilo and arrive wasted to camp for the start of the work week'*. He also says that *'As an ersatz Englishman I was always proud that the UKIRT people never put on airs. In contrast, for whatever reason, some of the other astronomers who came from around the world, including the University of Hawaii, would come over and expect special treatment because they were a Ph.D. etc. Terry Lee and Terry Purkins would sit around drinking just like one of the guys. In addition, UKIRT people would always tell us to help ourselves to their beer. I liked to think that it was because they understood how the world worked. A sandwich made by a cook who likes you is always far superior to a sandwich made by a cook who does not. I am proud to say that the UKIRT contingent stayed true to that principle in my five years on the mountain'*.

Initially the Canada-France-Hawaii project had its own dormitory and a separate building with its own kitchen which was also serviced by Hawaii Island Safaris. However by 1977, the Institute for Astronomy, who were ultimately responsible for supporting the mountain and who had failed to interest a private corporation in providing the services commercially, began to plan for the University of Hawaii to assume responsibility for the entire mid-level camp. The scheme was that the University would take over ownership of the existing buildings and refurbish them before operating the whole site on a not for profit basis on behalf of all the users. UKIRT management agreed with the scheme in principle provided they could be assured of access to the accommodation they would need (a baseline of 10 single rooms) and an equitable scheme for paying for them. Mindful of the fact that the University of Hawaii would be both the operator of the facility and at the same time a user of it, Vincent Reddish insisted on the setting up of a user committee to ensure that the management of the facility would be subject to regular consultations with all its users.

So in 1978 the University of Hawaii established the Mauna Kea Support Services unit (MKSS) as a quasi-independent group within the university to run the mid-level facility. The negotiations over costs, ownership and occupation rights of the existing buildings were, however, quite protracted. For example, Colin Humphries argued for delaying the change-over until construction was finished in order to retain control over the number of rooms available to his team in the event that unexpected demands for extra staff on the site should arise. A further stumbling point concerned accommodation for staff connected with the NASA IRTF telescope. Renovating the existing CFHT and UKIRT buildings to provide more spacious individual rooms would inevitability cost money and reduce the number of people who could be accommodated, and the available space was virtually all required by UKIRT and CFHT staff. Accommodating the IRTF team, who had not been put to the expense of having its own building (although it is questionable if IRTF would have been allowed to build one by the planning authorities), would require retaining at least some cubicles with bunk beds in either the UKIRT or CFHT building. As a consequence of this, and other detailed discussions over the costs and payment schedule for the refurbishment of the UKIRT building, the CFHT dormitory was converted first. Conversion of the UKIRT building was scheduled for the spring of 1979.

Eventually all the dormitories were turned over to MKSS who have managed the buildings and lodging arrangements ever since. As part of the new arrangements the UKIRT kitchen was shut down and the CFHT kitchen was used by MKSS to provide meals for staff from all the telescopes. John Morris transferred to the new organisation and worked with Chef Jimmy Nojiri, who would later manage the new facilities when they were eventually built. The new arrangements allowed the existing rooms to be gradually improved, although toilets and showers were still shared, and provided a multi-function building which served as kitchen, cafeteria and provided a recreational area which boasted a pool table, dartboard and a TV set. Alan Pickup (who had joined ROE in 1968 and had been involved in satellite tracking, site testing and a few other projects, arrived in Hawaii in 1981 to join the growing software team), remembers that the TV *'Could just about receive one channel'*. Perhaps because of the poor TV reception the pool table was very popular with astronomers, support staff and visitors. Patrick Moore thrashed all comers while visiting in 1981 and on another occasion one member of the Japanese Society for the Promotion of Science gave a very good account of himself.

**Fig. 4.10** Hale Pohaku common room and library in the late 1970s (Photo Ian Robson)

As the construction work finished the huts were converted to an all single room arrangement for the longer term. Each had a bed, wardrobe and table. The wardrobes had wire coat hangers in them for the temporary guest, which turned out to have a quite unexpected secondary function. If there was any earth tremor they were an excellent earthquake detector. Alan Pickup recalls that *'From time to time you would be lying in bed and then there would be this rattle, rattle, rattle and you knew there was an earth tremor going on'.*

Despite the difficulties of staffing and supporting such a remote location the cuisine at Hale Pohaku was reportedly reasonable (indeed a medical summary would describe it as excellent and varied) but it was a multi-cultural environment and there were different reactions to the food. For example, it took a little time for the issue of rice versus potatoes to be solved by providing both. Terry Lee remembers that the food could be first class. He says that *'Bob Qeen produced an excellent duck a l'orange for an IRTF event. He offered me the leftovers for breakfast after a night's observing, and champagne to wash it down with'.* Even so from time to time the Edinburgh team would attempt to supplement their rations with something more familiar. On one occasion John Clark had shipped over a tinned haggis, probably for Burns night and the Scottish contingent were all looking forward to this rarity. Tragically, when the can was opened it was found that it had a pinhole and the haggis had to be thrown away. It was a big disappointment.

The privations of these early huts have taken on somewhat legendary proportions and Ian Robson is happy to recount his memory of the sign saying "Save water, do not flush toilet". Richard Hills, one of the early sub-millimetre observers recalls that there was no washing machine and that he once drove down with a team of 6 people who had been up for 10 or so nights. As the vehicle full of dirty laundry descended into the warm and humid air towards sea level its interior became steadily more odoriferous. It is true the buildings had little in the way of heating or air conditioning and that in the very early days the mix of day and night workers meant that one group's activities would inevitably disturb the other's rest. However, in fact they were perhaps not as bad as all that. Gillian Wright remembers them as being rather better than the very basic facilities provided for UK observers at the smaller IRFC in Tenerife.

There was however an issue with provision of accommodation for female astronomers, who were fairly rare in those days, and there were no separate women's facilities. The huts had a row of rooms with beds in them and at either end of the corridor was a bathroom with toilets and showers, but there was no "ladies block" so to speak. Gillian remembers one observing trip when she and Norna Robertson, who was a post-doc at Imperial College, were going observing and Bob Joseph got a message saying that they couldn't stay at Hale Pohaku because they did not have space to make women's accommodation. In practice what making women's accommodation actually meant was assigning any female visitors rooms at the end of a block next to one of the bathrooms and then sticking a temporary notice on the door to say "Women's shower room". The problem was that by that time Hale Pohaku was getting very busy with lots of observers coming out so it meant that they had to shuffle all the men around so that for the nights that

Gillian and Norna would be there, so that there would be a suitable block of rooms. It didn't stop them from going in the end, but Gillian remembers that there was one step of extra hassle involved in getting women's rooms.

These temporary buildings were eventually torn down in 1991 – 8 years after the new mid-level complex, which is described later, had finally been completed.

## Sea Level Base

Notwithstanding Roger Griffin's one-man crusade to build a residence at the summit, and an early expectation that most of the team would live at the mid-level facility so they could be on-call at night, the logic for a base facility at sea level seems to have been gradually accepted. However, the location and size of the permanent sea level base remained an open issue for some time. The choice quickly came down to two possibilities, Hilo or the small cattle ranching town of Waimea, some way inland.

In November 1977 Harry Atkinson from the SRC office in London had visited Hawaii and Alex McLachlan arranged a meeting between him and the local councillors in Hilo. This took place in one of the hotels in Banyan drive. Alex remembers that '*The local council were a bit concerned about the developments on Mauna Kea. They had been very upset by the approach of the University of Hawaii people when they built the first telescopes there and made that scar on the mountainside (ie the road). They had upset a lot of people. One day I was sitting beside a swimming pool at the YWCA and a woman beside me, realising who I was, had commented, "Please don't spoil our mountain"*'. The following day Alex took Harry Atkinson to Waimea and left him to stay overnight at the Waimea Inn. The two men met a couple of years or so later and Atkinson asked '*Do you remember the day you dropped me off at Waimea? It was like that film with Spencer Tracy called Bad Day at Black Rock, when he got off the train and looked around*'. Alex says he understood exactly what Harry meant!

Despite the town's small size and relative isolation, the CFHT teamed seem to have preferred Waimea because of the climate and the high quality of the privately run Hawaii Preparatory Academy School, which was based there. According to John Jefferies '*The [CFHT] Board met with Richard Smart (owner of the Parker Ranch which was the largest on the island) who indicated his willingness to make a piece of land available in the township. The Board wanted a donation of land, or at least the possibility of purchasing it on favourable terms, but Smart wanted an arrangement under which the land and improvements would revert to the Parker Ranch after about 20 years*'.

The original proposal being discussed by the CFHT and the University of Hawaii was that a core building containing central facilities would be constructed on land belonging to the Parker Ranch, with outlying extensions being added as required. These buildings would then be leased to the telescope owners, possibly for a nominal rent, which might increase to something more "commercial" after a set period. If other groups, such as UKIRT, could be persuaded to join the scheme then other buildings, specific to each organisation's needs, could be added. The CFHT board decided in July 1977 that they would indeed build their headquarters in Waimea and this plan was ratified at the end of the year despite a belated attempt by Hilo councilwoman Helene Hale and a group of local business leaders to bring CFHT to their side of the island. Although a third option, in the form of a Waimea site under the jurisdiction of the University of Hawaii, emerged in late 1977 the CFHT board eventually decided to buy land on the open market. The land they chose was adjacent to some state land, holding out the possibility that if the University of Hawaii built a support facility there, then the joint base camp options would still be feasible.[1]

The decision of the CFHT board to go to Waimea placed the UKIRT group, or more specifically their masters at SRC, under pressure to make a decision before they found themselves disadvantaged by whatever arrangements were put in place by CFHT. A significant factor was the desire, which was shared by the CFHT and the University of Hawaii, to have all the main users of Mauna Kea based in the same place and operating as equals. This latter issue continued to be a major concern for the British and French-Canadian groups. As with the mid-level facility, the University of Hawaii looked set to become both the operator of the sea level base facility and a user of it, in which case it would become a customer of itself with the inevitable issues over possible conflicts of interest. According to some the University looked as if it wished to be the "King of the Mountain" and this was not acceptable to the other groups. While the political debates continued, a report evaluating the two towns in terms of infrastructure, commuting distance, ease of international access and the ability to recruit and retain staff was written. This report, in the absence of financial considerations, came down in favour of Waimea.

The debate within the British camp rumbled on into 1978, with telexes and letters passing back and forth around the triangle of Hilo, Edinburgh and SRC central office. There were further meetings between Harry Atkinson, members of the CFHT board and John Jefferies who recalls *'Simultaneously the British were looking for a site for their UKIRT headquarters and, after some hesitation, decided on Waimea also. Their hesitancy arose because their Project Office had been set up initially in Hilo. The two groups (CFHT and UKIRT) purchased adjacent lots for their permanent buildings next to a State parcel where the University was to build its own base of operations for the Institute. It all seemed like a fine idea'*.

Terry Lee was involved in this land purchase. In 2010 he wrote *'When I arrived in Hilo in July of 1978 there had been some serious discussion with CFHT about land for adjacent base facilities in Waimea. CFHT had proposed theirs on their current site and us on a smaller site behind them. Colin (Humphries) did not like this and proposed that we should go for an adjacent site similar in size to the CFHT one'*. Terry also did not like the idea of being blatantly upstaged by the CFHT and

[1] The parcel proposed by Richard Smart was subsequently given to the Keck Observatory for its Big Island headquarters.

they put this to the SRC via Vincent Reddish at ROE. At this point, because the sea level base facility was going to be an ongoing concern for the instrumentation and operational phases well after the telescope construction had been completed, Humphries handed the matter over to Lee who proceeded with the details of the land purchase, working with local lawyer Bernice Littman in Honolulu and operating under remote supervision from the UK. It was not a trivial problem, the value of the land was such that authority to purchase it would be required from the very highest level, coming from the government department, via the SRC council and its Astronomy and Space Research board, to ROE and hence to Terry. Furthermore the CFHT and UKIRT teams had to be mindful of two other risks, the possibility that the University of Hawaii might never join the Waimea scheme and that the gradual descoping of the mid-level facility at Hale Pohaku would require correspondingly larger facilities lower down. So enough land would be needed for whatever common facilities the two remaining organisations wanted to build and share.

Approval for taking a financial option (essentially putting down a deposit on the land) for the scheme, subject to various safeguards, came from the Department of Education and Science in April 1978 although wrangling over how much land was needed, and what would be done with any excess land if the University of Hawaii did eventually build in Waimea continued. Progress was slow, with some legal issues about the land creating problems. Although details of the type and size of building that would be needed had been set out in March 1979 by May no deal had yet been done. Permission for an immediate and outright purchase of the land (rather than taking an option while details were being worked out) was given to SRC by the Department of Education and Science and by SRC headquarters to Terry in August. A deposit towards the purchase of 2.27 acres for a price of $211,000 was finally paid on 7 August 1979, but this was not to be the final chapter of the saga.

Despite the growing expectation that the UKIRT base would eventually be in Waimea, it seems that Hilo was favoured by many of the UK team for pragmatic reasons, i.e. the proximity to the airport, the port where so much equipment would be arriving and the availability of local engineering infrastructure. Furthermore Hilo had ten times the population of Waimea, the island's main hospital, a range of affordable housing and several schools. This view was noted as early as 1977 by Harry Atkinson who reported it to the SRC chairman in 1978, although he also commented that there was a general feeling that future development would be centred on Waimea. Similar issues about staff preferences for Hilo were expressed to ROE management who noted the split of opinions and the fact that no solution would please everyone. To Alex McLachlan the question was not so much *'Why did UKIRT stay in Hilo, but rather why did CFHT go to Waimea?'*. In this regard he says *'To me this was rather odd. I could not understand why they were doing this. Hilo seemed to have all the necessary facilities required for importing the goods associated with the telescope, it had schools and the hospital. All the facilities were in Hilo. So I went to Waimea and look to see what was there. I passed this information back to people at ROE. In the end it seems that the CFHT did not have an economic or logistical reason for going to Waimea, it was purely cultural'*.

The Steering Committee debated the choice for the sea-level HQ and Ian Robson was insistent that Hilo was the only suitable site. He based this conclusion on the fact that he had observed on Mauna Kea, using the 88 inch telescope twice before and on his fact-finding visit on behalf of the Steering Committee. During these visits he had become familiar with the Big Island and the transportation, communications, commerce, housing and university focus of Hilo compared with Waimea, which he felt was really in the backwaters. The committee agreed that Hilo should be the preferred site for the sea-level base and passed this advice up the management chain.

Richard Jameson remembers that there were *'Many discussions over the relative merits of Waimea versus Hilo. Waimea was considered to be a cold, damp and boring town with no social life. So Hilo won the day, despite advantages of sharing things like a library with CFHT'*. Bob Joseph also remembers that when he visited Hawaii for a conference on infrared astronomy in 1980 Andy Longmore and Peredur Williams took him to lunch and argued strongly for a base in Hilo, chiefly on the basis that Hilo was much more desirable for people with families and the availability of resources for them. According to John Jefferies eventually *'there was a change in the directorship at ROE and the new appointee did not wish to overrule the strong preferences of his staff – by then happily settled in Hilo. So, for better or worse, we finished up with two sites for the operational bases'*. At that time the project team was installed at Leilani Street in Hilo and while the matter of a permanent base was a consideration, the priority was to get the telescope up and running and to build up the operations staff.

Not just the location, but also the size of the base facility was still being debated. On the 8th of May 1978 Roger Griffin wrote to Jim Ring saying that any base camp should not go to Waimea just because that was where CFHT was going. He also argued against a plush and well equipped overseas astronomical centre built with UK money, which would drain away talent from the UK itself. His concern was shared by others who did not want to see a repeat of the situation in Australia where the Anglo-Australian Observatory headquarters and laboratories in Epping had been built on a large scale and had become a major astronomical centre in its own right. The fear was that an overlarge Hawaii headquarters for UKIRT might prove very attractive proposition and allow resident astronomers in Hawaii to gain most of the observing time at the expense of the UK community they were there to support.

While discussions were being held about the merits of Hilo and Waimea, Roger Griffin continued to argue for living quarters and laboratory space at the summit, rather than at Hale Pohaku which he felt was far too low for effective acclimatisation. This debate continued with a letter to Vincent Reddish on the 1st of September 1978 asking for copies of the medical evidence on acclimatisation that argued against his view that a sea level base was impractical. Reddish replied that, whatever the medical issues, planning and logistical constraints meant that there were no practicable alternatives to having a sea level base. This correspondence continued into 1979 culminating in a statement from Griffin that, until the available medical evidence was circulated and examined objectively, he could be of no further service to the committee.

Ray Wolstencroft begged to differ with Roget Griffin's views since he also had direct experience of the issue, in 1961 and 1962 he had spent two, four month long, seasons working at an observing site in the Bolivian Andes at an altitude of 17,200 ft (5240 m). During the first season he had alternated 2 weeks on the mountain followed by 2 weeks rest at La Paz, some 5000 ft (1500 m) lower. The following year his group had commuted nightly to the summit from La Paz. He found that, despite the extra fatigue of the daily driving on poor roads, the latter regime was much more efficient with fewer problems from headaches and nausea and a general tendency to find problems at the site much easier to identify and think through. In May of 1979 Ray wrote a paper to the UKIRT Steering Committee entitled "Should the Support Facilities for UKIRT be placed at the summit of Mauna Kea?" In this paper he discussed the medical and logistical issues at length but also noted that a request by the University of Vienna to build a 1 m telescope on the summit had been stymied by the requirement to remove an existing telescope. Although this latter point may not have been the whole story of the Austrian telescope (the plan was to replace an existing 24 inch (60 cm) telescope and no new building permits would have been required) it was symptomatic of the general concerns about building on Mauna Kea. About the same time (1st of May 1979) John Jefferies wrote to Ray saying that the building of living quarters at the summit would contravene the spirit if not the letter of the lease granted for the telescope building. Furthermore, he warned that any attempt to modify the lease to allow such a building would have zero chance of success and would have a large and negative public relations impact.

Griffin responded to the paper by Ray Wolstencroft with one of his own. In this he restated his arguments that for proper acclimatisation and efficiency, the astronomers should remain at the summit for long periods and that the arguments being used against this view were not supported by the facts. He argued that the case for a sea level base was being made based on bureaucratic and administrative arguments to do with the welfare and convenience of the staff, not the efficiency of the observers. In addition to an annex which contains a point-by-point rebuttal of the Wolstencroft document he included a spoof paper showing that most of Ray's objections could be solved by building the telescope in Swindon gas-works using a converted gasometer in place of a dome! Malcolm Longair, by then ROE director, was amused by this latter contribution.

Roger Griffin would fight a losing rearguard action for some time yet, for example on the 30th of April 1980 the UKIRT Steering Committee were invited to comment on a case being put forward by Terry Lee for a computer at the UKIRT base facility. If support was forthcoming from this poll of the committee then it was proposed to make a case to an AII committee meeting on 17 June 1980, but if necessary, a special meeting of the Steering Committee could be scheduled to discuss the issue. In keeping with his views that money should not be used to build up facilities at Hale Pohaku or sea level in preference to at the summit, Roger Griffin objected to the proposal, fearing that if a computer was available at sea level it would free Terry Lee from any need to go to the telescope itself. In retrospect Terry finds these concerns rather amusing, he says *'Certainly a computer in Hilo would never have taken me away from the mountain top. I think I spent more time there than any of my successors. The two main arguments for the computer in Hilo were to make system development more effective by enabling the software guys to work in Hilo and to enable visiting astronomers to review their data before going home'.*

Although it had been clear to the UKIRT Steering Committee and to many of the UKIRT staff for some time that Hilo was going to be the most likely outcome, the debate over the location of UKIRT's base facility was finally ended soon after Malcolm Longair became the new Director of the ROE in late 1980. In his view *'The most important thing was the reluctance of the staff to move to Waimea for good logistic and societal reasons. I gave very high weight to ensuring that the support staff were happy and understood that we were listening to their views and looking after their interests. Secondly, the logistics were clearly better in Hilo because of the port and the airport. It was also the centre of social and political activity on the Big Island and we had established excellent relations with the Hilo authorities. Fourthly, I saw the possibility of developing closer relations with the University of Hawaii and interacting with the Hilo campus for the training of support staff and technicians. I was also persuaded that the idea of a science campus on the Hilo site would be the way to the future'.* Malcolm's decision, which was known to staff in Hawaii by early 1982, did however create one loose end. By now SRC had completed the purchase of the plot of land in Waimea, land which it no longer needed. After someone at the CFHT had told Terry Lee of a potential buyer, the SRC finance officer Lewis Addison entered formal negotiations to sell the land. The plot was eventually sold at a profit, which was duly returned to the government coffers, a rather unusual occurrence for a department whose normal function was to distribute money, not collect it.

# Chapter 5
# Making It Work

## Telescope Computer

By late 1978 the construction of the telescope was finishing and things were moving into the commissioning phase. The computer systems were primitive by modern standards, but they were just about as good as you could get at the time. The telescope was controlled by a dedicated PDP 11/40 computer housed in two large racks. The PDP 11/40 had a 64K word memory and two RK05 removable disk cartridges; basically giant floppy disks about the diameter of a pizza encased in a rigid white plastic housing about an inch and a half (4 cm) high. Each of these cartridges had a capacity of 2.4 Mbytes. The telescope software, written by Bryan Bell, was tightly structured and rather complex and, while it did a good job of controlling the hardware in an engineering sense, it had some shortcomings as an observational tool. For example, the telescope software understood standard astronomical definitions like Right Ascension and Declination, but had no concept of real astronomical coordinates on a given date, so finding stars was difficult! The solution was to use a program written in Basic on a Commodore PET desktop computer (or sometimes a programmable calculator) to convert between mean and apparent positions for a given time and then to put the resulting coordinates into the telescope computer. Some users even brought their own computers or programmable calculators to do these calculations themselves.

A second PDP/11 was used as an instrument computer. Its peripherals included a reel-to-reel magnetic-tape drive (for visiting astronomers to record and take away their data), a paper-tape reader and a Tektronix 4010 graphics terminal. The



J.K. Davies, *The Life Story of an Infrared Telescope*, Springer Praxis Books,
DOI 10.1007/978-3-319-23579-0_5

computers could communicate through CAMAC modules, which were used to interface with the telescope and the instruments. The telescope computer alone had no way of logging various parameters so the early pointing tests were rather laborious with coordinates, deviations, time etc. being logged by hand. There was also a teletype-like device called a Decwriter into which observers could type commands and which in turn recorded the instrument outputs so the astronomers would leave the telescope clutching large piles of line printer output.

Malcolm Stewart was one of the first generation of software engineers who had to support the system. He had started working on instrumentation in Edinburgh in 1976, developing software for the first generation of single detector photometers, CVF spectrometers and polarimeters. He arrived in Hawaii in the summer of 1978, got off the plane and went to his room at the Hilo Lagoon Hotel. The next morning, armed with a map of Hilo, he walked to Leilani Street, finding out in the process that American streets, even in Hawaii, are considerably longer than the ones he was used to. Malcolm remembers that 'To *boot the computer, one had to enter manually the bootstrap program using front panel keys. The bootstrap program comprised 10 to 20 words, each coded as 6 octal digits. Eventually I knew this by heart and could key it in double quick fashion'*.

The RT11 operating system was very simple, as indeed it had to be with so little available memory. It was a single user system that allowed two tasks to be run, one in the foreground and one in the background. Like all computer hardware, care had to be taken to ensure that the PDPs would operate at high altitude and would not fail due to overheating or to disk heads having insufficient air cushion to avoid crashes. There were other problems of computing at the summit, the extreme dryness of the atmosphere, combined with artificial fabrics in the clothes the astronomers wore, often created build ups of static electricity that could earth through the electronics with predictably disastrous effects. Ray Wolstencroft had one such experience, *'One night a spark took down the computer program of our polarimeter. It was a thousand lines, so we just went down to Hale Pohaku and retyped the whole thing. We only lost a night in the end'*. Peredur Williams had a low tech solution to this problem. He spent most nights at the summit boiling a kettle to keep the control room humid enough to stop the computer crashing.

**Fig. 5.1** The original control console was a clunky-looking setup aptly called "the battleship" because of its boxy appearance and grey colour. This is how it looked in the late 1979/early 1980 period. The squares drawn on TV mark the infrared beams, two because they were using focal plane chopping at the f/9 focus , also seen are the lock-in amplifier, a paper strip chart plus the TI59 calculator and printer to right of the operator's chair for getting apparent places, precession etc. The two meters near the TV were for monitoring the current going into the drive motors in case they overloaded (Photo Peredur Williams)

## A Slip-up with the Mirror

In the autumn of 1978 there was a set-back that hadn't been envisaged. During routine testing of the telescope control system the telescope was tilted over too far towards the horizontal. This resulted in the primary mirror moving forward so that the three stainless steel pads cemented to the inside bore of the mirror central hole slipped off the radial position-defining arms of the mirror cell, and were no longer in contact with them. This was potentially disastrous because a wrong decision about how to correct the situation could have stressed the mirror and endangered it. So, for the moment, the telescope was left pointing in the direction at which the incident occurred.

A staff meeting at Hale Pohaku was immediately arranged and, Colin Humphries says '*Rightly or wrongly, I (together with Terry Purkins) decided that nothing was to be gained for UKIRT by discussion about this with staff of the other telescope groups at Mauna Kea. Therefore an embargo was put in place on mentioning the subject outside the project group itself. Instead, calls were put through to the UK to request the presence on-site as soon as possible of Denis*

*Walshaw and David Brown so that properly considered corrective action would be taken. They arrived in Hawaii within 3–4 days and the mirror was restored safely to its normal position by carefully applying forces to appropriate outer radial levers and gradually changing the telescope pointing direction back towards the zenith. In retrospect there was probably no real benefit [in] placing an embargo on mentioning the incident to those in other groups (it was not a popular decision among the UKIRT staff) but it did underline the potential seriousness of the situation and the need not to make hasty decisions, particularly when working at high altitude'.*

In November 1978 ROE director Vincent Reddish visited Hawaii to see how things were progressing and although he seemed pleased with the progress that had been made with the telescope, there were some staff issues that had surfaced during the visit which, he decided, needed to be rectified. So, on his return to Edinburgh, he came directly to the Observatory and called a number of people to his office, asking them to get on a plane to Hawaii as soon as possible with a view to boosting the size of the team. Within a few days Andy Longmore and Peredur Williams were en-route, arriving in Hilo on the 1st of December 1978 and returning to the UK for Christmas. Andy and Peredur worked on telescope commissioning, reporting that the pointing and tracking performance was not yet acceptable. Pointing tests, and development of software to reduce the errors, would continue well into 1979. On the technical side Ralph Martin and Hamish McFarlane were also sent out, with Ralph working on the computers and Hamish on problems with the sticking dome.

At roughly the same time Ian Sheffield and Bill Parker were earmarked for return to the UK, although in the event only Ian actually left. Ian attributes his being *persona non grata* to a specific incident in which *'Bill Parker and I got so tired that one evening when we were supposed to install and test the autoguider we set off up the mountain and when we got about half way we realised that we had left the autoguider back in Leilani St. We were so tired we said "Oh sod it, we're going home", which we did. All hell broke out the next day because we had wasted a night on the telescope'*. Bill Parker does not remember giving up and going home but agrees that *'it was seen as a negative incident and certainly prompted our potential return'*.

Ian concedes that since the telescope engineering time was very tightly scheduled and there was pressure to get the telescope into operation they should not have done what they did, but in reality it was more complicated than this. A small team of often strong minded people were under pressure to deliver and there were undoubtedly tensions developing amongst them. Colin Humphries was more of a scientist than a manager and Terry Purkins was a hard driving, ambitious individual with a strong sense of getting the job done. To them anything less than 100 % effort was not enough and Terry Purkins at least regarded Ian's unofficial work at Radio Shack as showing a lack of commitment to the job at hand. To Ian however the recall was *'a complete shock to Bill Parker and I. It came straight out of the blue'*. Bill Parker elicited the help of his trades union and secured a continuation of his service until the 1st of January 1986 when he left to live and work in California. Nonetheless he recalls, on more than one occasion, finding himself *'boiling with rage'* after speaking with ROE and local management.

During the whole of 1978 which included the erection of the telescope in Hawaii, installation of the optics, and the commissioning, Gordon Adam at ROE had been co-ordinating the actions that were required at the Edinburgh end. This involved maintaining contact with UK contractors, smoothing arrangements for staff that were about to be transferred to Hawaii, and dealing with a host of technical matters or queries that were brought to his attention from Hawaii. There were usually daily exchanges of telexes between the members of the team in Hawaii and Gordon in Edinburgh.

## Astronomer-In-Charge

By the time Peredur Williams returned to Hawaii in early in 1979 for a 3 month long visit, the main engineering work on the telescope was complete and Terry Lee had taken over the local management. Interestingly Terry was described as the "Astronomer", rather than the "Officer", in Charge giving him the moral authority to do research as well as management. The transition took place on the 1st of January 1979 and according to Colin Humphries it was more than a notional handover, he remembers that Terry Lee and David Beattie *'Were up at the telescope at 00.00 hours on 1 January 1979 standing around waiting to take over, whilst Terry Purkins and I had been up on the top of the dome at midnight that night in freezing weather trying to crack ice off the dome shutters in order to complete what we wanted to do'*.

Lee and Beattie had a very different style of management. According to one staff member *'Terry was hands on in terms of the technical stuff but he very much let people manage their own things, he steered the ship rather than trying to manage it'*. David Beattie had been involved the ROE rocket programme and, according to some who knew him, was *'A tremendous motivator and man manager. Terry [Lee] was very bright, but was not a good communicator, he just expected you to read his mind about what he wanted you to do. If you could read it, good, if not, well...... David was the facilitator'*. Some people feel that David Beattie was not given enough credit for what he did. Dave Robertson says that *'He led by example. He would not ask you to do something he would not do himself and that was something I learned from him. I might ask someone to do something I cannot do, but I won't ask people to sweep the floor or clean something down if I can do it'*. It seems that *'Terry was a thinker but not much of a communicator but David was a communicator, a bit larger than life perhaps, but certainly they worked well as a pair, as a partnership'*. Terry Lee agrees that Beattie was an exceptional person who contributed greatly to UKIRT. He says *'We worked closely together in a way I have never worked with anyone else and the whole was much greater than the parts. It was not easy on either of us when the spell was broken, but especially for David who remained in Hawaii perhaps under, rather than part of, [the] new management'*.

However, if Terry Lee was working well with his team in Hawaii, relations with higher management at Edinburgh were not always so smooth. As the Astronomer-

In-Charge in Hawaii, Terry was subordinate to Ray Wolstencroft, the head of the UKIRT unit at ROE, rather than reporting directly to Vincent Reddish, who was then still the ROE director. Unfortunately this arrangement did not always work well and in 1979 there were exchanges between Ray Wolstencroft and Terry Lee on the division of responsibility regarding the delivery and commissioning of instrumentation for UKIRT. Ray wanted ROE to deliver only instruments that were already fully operational and merely required commissioning in Hawaii, arguing that the staff in Hawaii was too small to carry out laboratory testing of partly finished instruments. He says that, although he was aware of the pressure they were under, he was frustrated by the lack of information coming from Hawaii back to Edinburgh. Terry felt the same way about the lack information on progress of the instruments and delivery timescales coming from Edinburgh. Vincent Reddish became involved, expressing his concerns about this apparent failure of communications and was very firm in wanting any communications from Hawaii to SRC headquarters in Swindon to be approved by ROE in order to maintain proper management controls and presumably to protect the long term interests of the ROE. He was obliged to repeat his insistence that Terry not communicate directly with Swindon over issues of staffing or budget following an incident in which this had happened somewhat informally and without him being alerted.

**Fig. 5.2** David Beattie pops up through the UKIRT primary mirror (Photo Peter Forster)

Tension between ROE and Hawaii continued to surface from time to time, an example being Terry Lee's criticism that the initial arrangements for the planned dedication ceremony devalued the work that he and Colin Humphries had done in building and commissioning UKIRT and diminishing their stature in front of the local Hawaiian community. Nonetheless for the moment, the management situation remained unchanged with Ray as head of the UKIRT unit at ROE, Gordon Adam acting as his deputy, Richard Wade having responsibility for infrared instrumentation and Alan Pickup working on software. Also in the group was a young administrative assistant called Frances Shaw who was recruited in 1979 after *'Trudging up Observatory Road wondering what on earth someone with a degree in French was supposed to know about astronomy'*. She soon became the "UKIRT dogsbody", doing anything from recording and distributing telex communications to revamping the UKIRT filing system and making the tea.

## First Infrared Light

Once the telescope was working the operations team were able to start work on installing the infrastructure to support those scientific instruments which could be used at the f/9 focus. According to Terry Lee *'One of the first things we did was to put on a TV camera which wasn't very good in terms of sensitivity but it was {good} in terms of resolution. One of the things we saw from time to time were images that looked to be around half an arcsecond in diameter. I told this to Vince Reddish on the phone and he said "Well that can't be, something must be wrong with the TV camera. I wouldn't shout about having half arcsecond images or people will laugh at you" '*. In fact this concern was not well founded, the excellence of the site had already been demonstrated in 1971 when the CFHT site testing team had experienced a 6 week period of very good atmospheric conditions. Interestingly John Jefferies had had similar thoughts almost a decade earlier. In his unpublished account he had written *'Thus we found, for example, that our double-beam results (from a telescope a few feet above the ground and inside a windscreen) indicated seeing consistently around 1/3 of a second of arc; values three times as high were regarded as good seeing at established sites and I was hesitant to publish these numbers without further critical study. In particular I wanted to understand better how these double-beam measures would relate to the seeing that would be experienced at a real telescope at the same site. I was quite sure that our results, coming as they did from people without any background in site evaluation, would be regarded with great scepticism by the astronomical community'*

Later in January came the milestone of first infrared light. Dave Beattie prepared and Terry Lee mounted the UKT1 infrared photometer on the telescope. At dinner that evening in Hale Pohaku American infrared astronomer Eric Becklin said that he had heard that the UKIRT team were planning to get first IR light and asked if he could watch. The UKIRT astronomers agreed and the three of them went up to the telescope. Terry Lee recalls that the conversation went something like this:

Becklin: "What size aperture have you got"
Beattie: "5 arcseconds",

Becklin: "Do you have a bigger one "
Beattie: "Yes"
Becklin "Are you going to use it?"
Lee and Beattie in unison: "No, we don't need it"

Back in the control room they set the telescope on the target, centred it on the TV and immediately had signal. Terry Lee remembers that *'Even Eric was a little bit impressed, and he is a hard man to impress'.*

A few days later, on 28 January 1979, Peredur Williams took the first spectrum with UKIRT, a 3 μm observation of the star EZ CMa using a CVF in UKT1.

After these achievements work continued to try and improve the pointing and tracking issues still affecting the telescope and to fully commission the first two scientific instruments, the f/9 photometers UKT1 and UKT2 along with their focal plane chopper. These instruments both boasted an InSb detector with JHKL and M filters and a CVF. The CVF's covered slightly different wavelength ranges, 2.3–4.6 μm in the case of UKT2. Also occupying the time of the commissioning team was the f/20 coudé focus, this was because the f/20 secondary could be installed without recourse to changing the top-end, an operation which was impossible while the dome crane remained hors de combat.

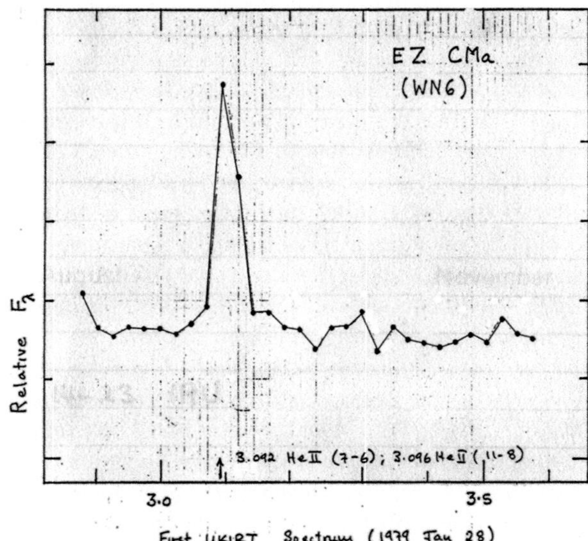


**Fig. 5.3** The first of many infrared spectra taken at UKIRT (Peredur Williams)

## Record Breakers

One morning, after the main UKIRT team had been in Hawaii for about a year, they received a visit from Alan Russell, the producer of the BBC children's programme "Record Breakers". He was in Hawaii because Hawaii has the world's largest mountain (Mauna Loa), the best surfing, the highest macadamia nut production, the wettest spot on earth (Kauai) and he had just discovered that Hawaii also had the world's biggest infrared telescope. Better still, it was British! He asked if they could include UKIRT in the Hawaii programme and of course it was agreed. A month later, in June of 1979, he came back with a cameraman and a sound recordist and the well-known entertainer Roy Castle, who presented the show. It was agreed to have an interview between Roy Castle and Colin Humphries on the telescope yoke. The producer's only worry was that Roy Castle was prone to forget his lines and since he often could not remember his script at sea level the filming could be a disaster at 14,000 ft. However they decided to give it a go and according to Terry Purkins it went perfectly first time. He says *'Roy Castle was excellent; so was Colin'*. As well as the few minutes of TV which was broadcast, the visit of the "Record Breakers" team was also recorded in one of the books of the series produced by the BBC.

**Fig. 5.4** Colin Humphries and Roy Castle on the UKIRT yoke filming a spot for the BBC programme "Record Breakers" (Photo Terry Purkins)

## Earthquake

One of the many of the elegant features of the UKIRT design was Denis Walshaw's system for earthquake protection. The design is such that the gravitational load of the telescope is transmitted to the ground through the north and south support columns which carry the weight of the entire telescope. Each of the support columns rests on a set of three large ball bearing races. The normal operating position of the telescope is defined by two 6 mm diameter brass pins that slot down through holes in steel base plates attached to each of the columns and into plates attached to the concrete foundations of the telescope building. These "shear pins" are machined down in diameter to 4 mm at their midpoints. The telescope was specified to be able to withstand earthquake forces of 0.3 g horizontal and 0.1 g vertical (an event of about 4 on the Richter scale) and in the event of the horizontal force being exceeded, the brass pins are designed to shear at their narrowest point and the momentum gained by the telescope is then dissipated by the whole telescope structure rolling on the ball bearings until its end stops are reached. The first such event occurred during the construction phase in mid-1978 when there was a small earthquake of Richter magnitude 4.2. The protection system operated as designed and afterwards it was just necessary to return the telescope to its original position using the built in screw jack and insert new shear pins. On the 6th of March 1979 there was another earthquake of Richter magnitude 4.7–5.1 which was again sufficiently powerful to break the shear pins. When the tremor subsided the telescope had moved about 4 mm and it was duly screwed back into its original position after which new pins were installed and the telescope pointing rechecked. On the 21st of September the telescope was again moved by an earthquake (Richter magnitude 5.5) but once again the shear pins broke and no harm was done.

With Hawaii being a volcanically active region, earthquakes are not uncommon and most of the UKIRT staff has experienced one or more. The first indication is often that the telescope suddenly loses sight of its guide star which has apparently flown off the TV screen. Then, once everything settles down it is usually a fairly routine task to check the shear pins and reset the telescope if required. Usually, but not always. After one earthquake Kent Tsutsui, one of UKIRT's locally recruited technical staff, was in the basement working under the telescope to set things straight when there was another "quake". He says *'I think I broke a record sliding out from under the telescope and down the telescope support pillar ladder. My heart was pounding like crazy. It 's true what they say about seeing your life pass in front of you when you think you are going to die. I saw a glimpse of mine. It wasn't the best of experiences'.*

## Starting Operations

During 1979 the scientific staff at UKIRT expanded to include four staff astronomers. Peredur Williams and Andy Longmore, who were by now on overseas tours from ROE, were soon joined by Bill Zealey who had been working at the UK Schmidt telescope in Australia. Bill's contract to work at the Schmidt was coming to an end and he was offered the opportunity of applying for a job back at ROE on the COSMOS measuring machine or trying for a position at UKIRT. He picked the UKIRT option and flew to the UK straight after an observing run. He arrived at ROE tired and jet-lagged just in time for morning coffee only to be told that his interview had been brought forward since a number of candidates had not turned up. He got the job anyway. Another arrival was Ian Gatley, who is seen in the photograph of the Leilani St offices. He had taken a degree at Imperial College in London before moving to the California Institute of Technology to do a PhD in infrared astronomy and was recruited to the UKIRT staff by Terry Lee in view of this, then fairly rare, background. Ian would be at UKIRT until the mid-1980s. During these early months Dave Adams from the University of Leicester spent a 4 month sabbatical at UKIRT working on improving the pre-amplifiers for the InSb photometers.

Before the telescope was open to users, Ian Robson was asked by Terry Lee and David Beattie to try it out for a couple of weeks over Easter in 1979. Unfortunately, the weather was incredibly poor for the entire time, and no observing was possible. The lack of pointing and poor performance of the guide telescope was, however, well noted and made the need for a guiding photometer for the sub-millimetre essential. Ian remembers one night when he and David Beattie were sitting out the weather, but exercising the telescope's pointing and tracking when they heard an odd grinding noise from inside the dome. On going to check what was happening they discovered a set of steps under the telescope which were being squashed as UKIRT slowly moved towards the zenith. It was a salutary lesson on the risks of working at high altitude.

Although it had been hoped that a regular programme of observations by visiting astronomers selected by the Panel for Allocating Telescope Time (PATT) could begin by the summer of 1979, the resolution of various mechanical issues forced an engineering shutdown over the summer. One problem was that the dome did not rotate reliably because the circular ring at its base distorted when there were strong winds which caused some of the motor drives to lose contact with the base ring. The underlying reason for the problems are now unclear, actual winds may have been excess of the specification and local alterations to the dome had been made to replace the original castors with "railroad type" wheels. Maintenance or operational issues may also have contributed. Observa-Dome made some repairs and Colin Humphries entered into negotiations with them to try and improve the situation regarding the dome crane. However, before a mutually acceptable solution could be reached, the UKIRT team decided to seek an alternative approach. They worked with engineering architects in Honolulu and unilaterally agreed to a re-design of the dome and its crane. This would involve welding a massive steel girder to the inside of the dome and strengthening the crane. The contract was awarded to Brittain Steel, a company in Vancouver, British Columbia, who completed the task by August 1979 at a cost of $70,000. During the period when the extra welding and steel installation work were in progress (about 4 weeks), the telescope structure and mirror cell were shrouded, thus halting the programme of work at the summit.

There was however a small amount of science done that summer, when the need for the shutdown had become obvious, two of the projects awarded time by PATT for the 3rd quarter of 1979 were brought forward and Chris Impey and Doug Whittet used the common-user f/9 photometer for projects involving BL Lac objects and interstellar extinction studies respectively before the telescope was closed. The next official observing did not begin until the 4th quarter of 1979 with the first time being given, appropriately enough, to Jim Ring's group from ICST.

## Dedication

The formal dedication of UKIRT took place on the 10th of October 1979 in an event organised locally by Terry Purkins and Colin Humphries while at higher levels there was some debate as to the sort of person who would bring the most cachet to the event. If the intention was to have an impact with the local community, and raise the profile of the Big Island and the State of Hawaii itself, then a member of the British Royal family was considered to be a prime choice. On the other hand if the idea was to attract the attention of the astronomical community in Britain, then perhaps someone more associated with astronomy, Patrick Moore for instance, would be better. In the end the decision was to invite as the guest of honour HRH the Duke of Gloucester. Terry describes the Duke as *'a very nice man, well briefed, who asked very sensible questions!*. Susan Beattie, who had driven up some of the press, agrees that he was *'A very nice man and seemed a little nervous'*.

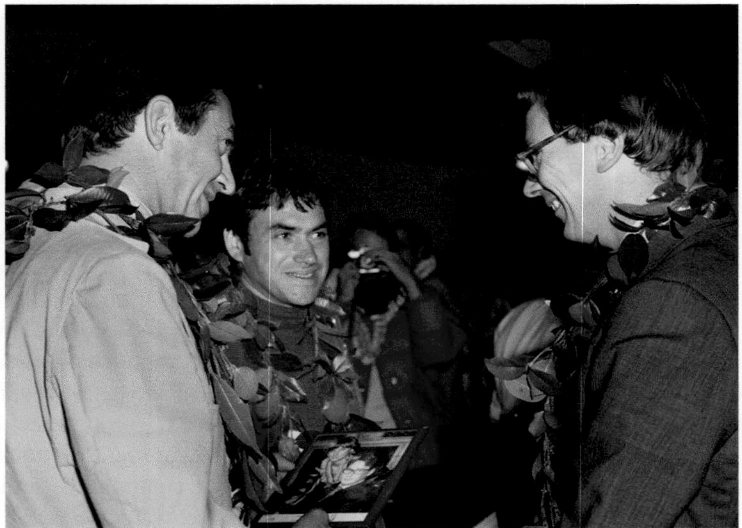

**Fig. 5.5** Colin Humphries, Terry Lee and HRH the Duke of Gloucester (Photo ROE)

Although there was light sleet at the summit, the ceremony was attended by about 90 people including scientists, contractors and local dignitaries. Also present was Sir Geoffrey Allen, the chairman of the SRC, senior managers from Dunford Hadfields and Gordon Carpenter's widow, Audrey. Jim Ring did not attend; he had a fear of flying.

There were only a few of the staff at the event. Dave Robertson describes himself as *'One of the privileged few who worked on the project who actually got to be there. I drove some dignitaries up. That was probably why I got to go'*. John Clark played the bagpipes, no mean feat in the thin air of Mauna Kea, but John says he had no difficulty and by all accounts the piping was a success. Terry Purkins describes the sound as *'Magnificent, even in the rarefied air of 14,000 feet'*. John cannot remember what he played, but does recall that *'The Duke of Gloucester's aide was Simon Bland, my ex-commanding officer in the Second Battalion Scots Guards where I had been his company piper. He obviously did not remember me after 10ish years but he made the point of introducing me to the Duke as his ex-company piper'*. There were speeches by Vincent Reddish and the Duke of Gloucester and a plaque was unveiled to honour Gordon Carpenter.

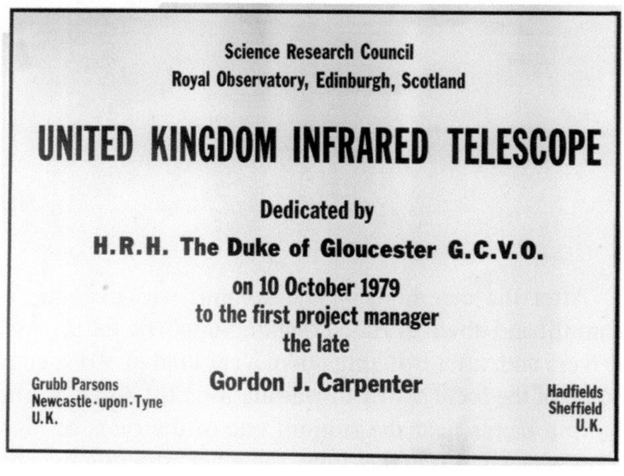

**Fig. 5.6** Memorial plaque in the UKIRT dome (Photo ROE)

**Fig. 5.7** Piper John Clark at the UKIRT dedication (Photo Terry Purkins)

After the ceremony at the summit was over the visitors were driven off the summit and down to Hale Pohaku. Support scientist Andy Longmore was one of the drivers and, after bringing down a car load of VIPs, he returned to UKIRT to collect some of the local staff still waiting for a lift down. Right on the middle corner of the hairpin bends near the summit one of the tyres of his vehicle blew out. Since the road was still dirt at that time, the wheel rim dug in and prevented him from turning to follow the bend. He stopped only a couple of feet from the edge of the road, narrowly missing going straight off and over the drop.

The day concluded with a dinner in a Hilo hotel. Amongst the guests were astronomers Richard Wade and Matt Mountain who had been scheduled on the telescope but were persuaded to close for the night in return for invitations to the dinner. Richard admits to having a serious night of drinking and, the next day, facing some challenging adjustments to their instrument while dealing with a hangover. John Jefferies remembers '*One other incident that comes to mind is the dedication of UKIRT, which was a blast. It was highlighted by a dinner with one of the Royals at a Hilo hotel with the champagne flowing freely. The head of the British government scientific research unit (SRC) was a contemporary of mine and knew all the scurrilous songs that I had learned as a member of the Cambridge University Rifle Association during our Bisley visits soon after World War II. He and I sang them together at the end of this very boozy affair to the wild applause of the half dozen bitter-enders – certainly including Terry Lee, the local Head of UKIRT with whom I enjoyed the best of relations. There was much Hawaiian entertainment there too – mainly due to Terry (I suspect) the UKIRT had excellent relations with the local community'.*

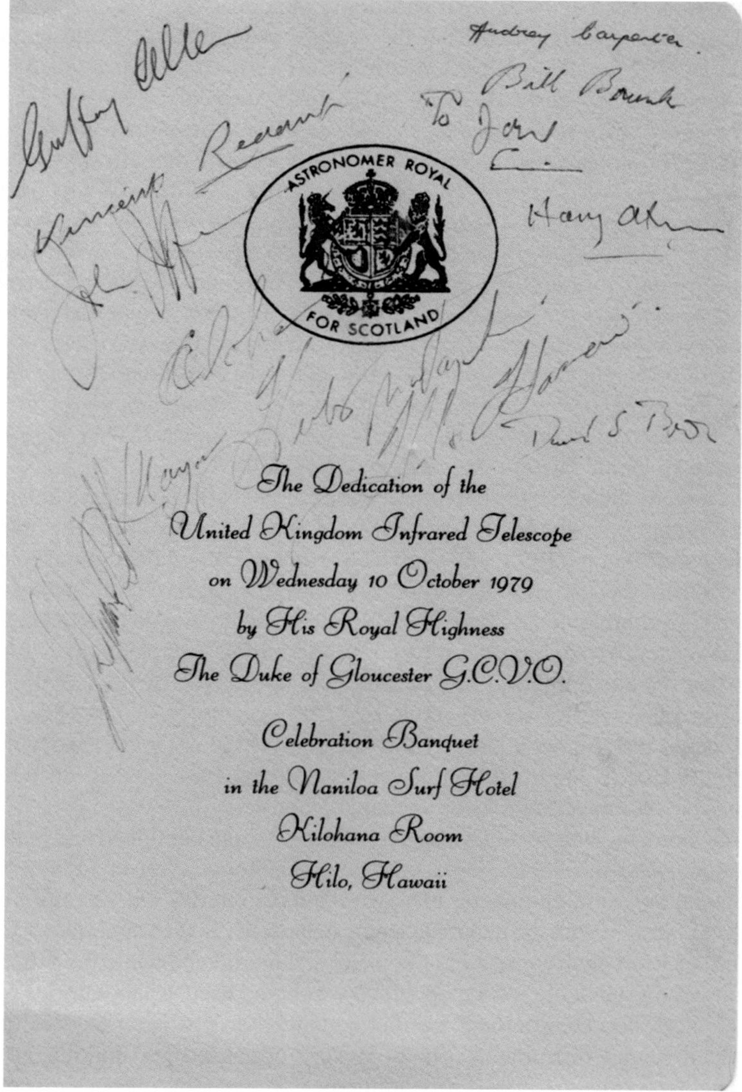

ASTRONOMER ROYAL FOR SCOTLAND

The Dedication of the
United Kingdom Infrared Telescope
on Wednesday 10 October 1979
by His Royal Highness
The Duke of Gloucester G.C.V.O.

Celebration Banquet
in the Naniloa Surf Hotel
Kilohana Room
Hilo, Hawaii

**Fig. 5.8** This menu was signed by a number of key people in the UKIRT project including Sir Geoffrey Allen (SRC Chairman), Vincent Reddish (ROE Director), Gordon Carpenter's widow Audrey, Bill Brunk (NASA liaison officer with the IfA in Honolulu), Harry Atkinson (SRC), John Jefferies (IfA), David Brown (Grubb Parsons) and the mayor of Hilo, Herb Matayoshi (Scan of original document provided by Susan Beattie)

According to Harry Atkinson *'The dress for the dinner was an Aloha shirt. A shop in the hotel sold them. One make was "HRH". We told the Duke and I saw him buying an HRH shirt that evening, quietly'*. Terry Lee also has an important memory of the event; in the early hours of the morning he was able to persuade Harry Atkinson to divulge who was to succeed Vincent Reddish as director of the ROE. At the time the answer meant little to him, his memories of Malcolm Longair were almost a decade old and the changeover was still a year away. In the meantime, with the dedication over, Colin Humphries and Terry Purkins returned to the UK in November 1979, their part of the job done.

Looking back on this phase from the vantage point of 2011 Colin Humphries recalled that *'The budget to cover the cost of the construction phase of UKIRT had been approved at an early stage by the UKIRT Steering Committee at £2.518 million, and the up-to-date spend was reviewed at each Steering Committee meeting. This figure included all contractual work in the UK and Hawaii, as well as travel costs. It excluded salaries and associated costs of SERC staff involved in the project. The small extra cost of the final primary mirror figuring was found from within the £2.518 million. Prior to the start of the installation phase I asked the Steering Committee to approve expenditure on the rental of suitable premises in Hilo to accommodate the staff working there and for associated costs such as the provision of vehicles for transportation between Hilo and Mauna Kea, etc. This was set at £50,000, which, together with the £50,000 for the contract to rectify the faults with the dome, brought the total budget for the construction phase to £2.618 million. The increased budget was approved and the entire UKIRT construction project was completed within the total of £2.618 million'*.

It was already clear that the handful of staff from the UK, there were only about a dozen, could not operate the facility unassisted and SRC rules about the number of permanent staff that the ROE was allowed to employ limited the options to expand this group from the UK side. Accordingly UKIRT began to recruit locally via the Research Corporation of the University of Hawaii since people employed via this route did not count against the allowed ROE staffing complement. During this process Lee remained mindful of advice he had been given by Sir Geoffrey Allen who had said that UKIRT should aim to be "*The best employer on the island*". Lee also discussed the still open issue of the location for the sea level base with Allen who advised that if good staff and good staff relations meant that, on reflection, plans to go to Waimea might have to change, then so be it.

For the moment, with so much else to do, the status quo continued and the UKIRT team remained in the, increasingly cramped, accommodation in Leilani St. The astronomers, software engineers, managers and secretaries worked upstairs in a large open plan area with the technical teams downstairs in the warehouse/workshop area. The sea level environment of Hilo is hot and humid and while the offices were nicely air conditioned, the workshop which was being used as laboratories was not. Since the workshop environment was not conducive to precision work on delicate and temperamental instruments, David Beattie organised the building of an air conditioned working space, a sort of miniature clean room, inside the warehouse within which the technical staff could work in more favourable conditions.

# Chapter 6
# Early Operations

## Pioneering Days

In the late 1970s infrared astronomy was a very new business, far from the computerised, automated and queue scheduled observing that is taken for granted today. Operations were so "lo-tech" by today's standards that even the installation of a telex machine in the summit building warranted a section in the primitive UKIRT newsletter of the day. The telescope had entered service without a chopping secondary so scientific observations had to be conducted at the f/9 Cassegrain focus using a mechanical focal plane chopper. This was done by fitting the instrument mounting system, known colloquially as the "Gold Dustbin" because of its cadmium plating, with a focal plane chopper supported on an arm that was pushed in via one of the six instrument mounting ports. There was not yet an intensified TV camera and the camera which was fitted had a poor limiting magnitude, making it hard to find guide stars, especially in the dark cloud regions UKIRT was intended to explore.

The oscillation in Right Ascension introduced by the telescope position encoders made it impossible to use small apertures for photometry so Bill Parker and others spent many hours trying to solve the problems of telescope pointing and cyclic encoder errors that affected the tracking. Originally it was hoped that the errors would be repeatable and could be taken out by applying corrections through software, but this proved not to be the case. It was clear that both the tracking and the pointing errors were unrepeatable. Parker recalls how the tracking problems were solved *'We were all trying to think how this could be so. After some months of thinking, it suddenly struck me that, as a consequence of the equatorial mount, once*



J.K. Davies, *The Life Story of an Infrared Telescope*, Springer Praxis Books,
DOI 10.1007/978-3-319-23579-0_6

*the telescope was pointing in the right direction it could be made to track simply by applying a constant sidereal rate in RA and Dec, the rate in Dec of course being zero. This was relatively simple to accomplish by using the encoders (the source of the problem) to point the telescope approximately, then with one switch and a couple of relays, to bypass the encoders completely and slave the telescope velocity to a constant value using only the output from the tachometers used to stabilize the servo loops. I used an Atari joystick to apply small increments and decrements in each axis to enable [fine] corrections. When an observation was finished, the astronomer would simply switch back to encoder control and the telescope would move to the (inaccurate) position given by the encoder feedback. This worked and gave confidence that the tracking problem really was in the encoder system. Later, when more accurate encoders were fitted, we found that the encoder couplings fitted by Dunford Hadfields had been badly overstressed and had in fact broken, but had still provided rotational information which was very poor in quality. In fact, while we were dismantling the old system, one of the encoder couplings literally came apart in my hands'.*

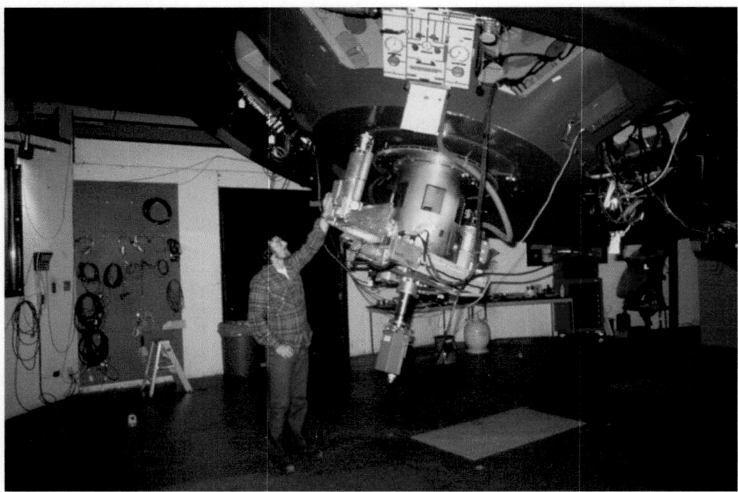

**Fig. 6.1** Ian Robson and the "Gold Dustbin". Two instruments can be seen on the left and right sides of the platform and a TV camera is positioned underneath to look up through the dichroic mirror (Photo Ian Robson)

As part of the efforts to solve the problems with the pointing of the telescope the engineers developed a programme of pointing tests to determine how the telescope was behaving and to ensure that it pointed as accurately as possible. These showed that the telescope structure suffered from mechanical hysteresis; that is to say the pointing characteristics differed depending on from which direction it approached a specific point in the sky. So there were eventually two sets of pointing references depending on where in the sky the telescope was looking and from which direction it had come to get there. The frequency of these pointing tests was high while the team were trying to figure out what was happening. As part of the engineering work to determine what was causing the hysteresis it was decided to install stress measuring devices over parts of the structure. This would normally have been a complex and time consuming operation but, knowing that Charlie Richardson was an experienced rock climber, David Beattie asked him to bring up his climbing harness and rope and help install them. Charlie says *'I thought he was joking at first but he was completely serious and for some weird reason I had indeed brought all my climbing equipment from the UK. I wish that someone had been there with a camera to see me hanging in my Don Whillans climbing sit-harness from various parts of the telescope structure as we rapidly installed the necessary equipment at key points around the structure. It was the only time I used the climbing equipment in Hawaii'!*

Bill Parker realised that the unrepeatability in pointing was due to the use of Belleville washers in the mirror supports. When the air was admitted to the mirror flotation bags, differing stiction in the washers would cause the mirror to point in slightly different directions each time. This was cured by fitting three Linear Variable Differential Transformers (LVDT) contacting the back of the mirror to measure the difference in mirror position and then correcting for the pointing errors in software. Bill thinks that it was Charlie Richardson who provided the software to reduce the LVDT outputs to pointing corrections and who later commented (with a grin) that had Bill used four LVDTs instead of three, his job would have been a lot easier.

However, there remained an additional element of unrepeatability in the pointing due to declination hysteresis which Andy Longmore and some of the engineering crew spent a lot of time investigating. The cause came to light when the mirror and cell were removed for aluminising. The screw jacks attaching the cell to the structure had not been tightened fully, presumably after the previous aluminisation. Alf Neild has an explanation for this. The second mirror aluminisation took place in the spring of 1980 and by then he was the only one remaining at UKIRT who had first hand experience of this operation. However, his first child was due about the time the aluminisation was planned so, as a precaution, Dave Beattie asked him to write a detailed Aluminising Process report in case he was suddenly called away. Alf found this very difficult to write, as trying to remember every detail of a very complex process, which he had only ever been involved with once and at altitude, was not the easiest thing to do. The document, UKIRT report number 11, said that the bolts should be tightened until '***most** weight* *is transferred onto lower truss members'*. It seems likely that different interpretations of the word "*most*" left the bolts too slack.

**Fig. 6.2** The crew involved in the second aluminising run described above. L-R Ken Maesato (Electrical), Charlie Richardson, Dave Beattie, Terry Lee, Sidney Arakaki, Alf Neild (Photo Alf Neild)

## Early Observing

There were, as yet, no night assistants (or telescope operators as they would later be known) so an observing team at the summit might comprise as many as six people. Between them they would; operate the telescope, control the instrument, review the incoming data, check the dome position (no automatic dome rotation system was yet installed), compute guide star offsets, carryout basic quality control checks and make the coffee. So it was quite a menagerie at times. However, gradually more of the more routine tasks were handed over to the computers, limited though they were in those days, and the telescope autoguiding system was improved.

The original autoguider for UKIRT was of an ROE design and was based on a vacuum tube photodiode in which electrons were liberated from a photocathode, accelerated and then focussed onto a set of four separate anodes in the form of quadrants of a square. It did not integrate the signals at all and so required a bright star to work properly. The autoguider was intended to be used with a piggy-back guide telescope, which in theory was aligned with the main beam, but differential flexure between the two telescopes meant this method was not a success. At the suggestion of Ian Gatley an intensified TV system was purchased for target acquisition purposes with a camera looking up through the dichroic mirror and displaying the image on a screen in the control room. This system provided the basis for an innovative, if rather "Heath-Robinson", autoguider. Bill Parker bought four silicon solar cells 10 $mm^2$ and had a diamond shaped holder with the minimum possible spacing made for them. This was then stuck with black electrical tape against the screen of the monitor. Some electronics were designed to accumulate the signals from each photocell over a single frame of the TV image, taking great care to keep the signals as linear as possible. At the end of each frame the accumulated solar cell signals from opposite pairs were taken to form an error signal for each axis and the error signals were fed into two servos.

David Robertson says that technically this was the most interesting thing he did. *'I did most of the work at the summit. I had a little electronics room which I built for myself in one of the closets and where I created a little bench. I was up with Ian Gatley to commission it and on the night when we went from hand guiding to this, the first thing it did was to shoot off. So I swapped the left and right plug over and this time it was straight in, locked. It was pure hardware, I had spent a few months building the system and, apart from one crossed wire, it worked first time. No-one was allowed to touch it after that. Ian Gatley would not let anyone go near it because it was the best guiding he had seen on the telescope since he got there. It was just stuck under the bench and when I was back in 1981–82, it was still sitting under the bench being used'*. This temporary autoguider worked surprisingly well and made a big difference to UKIRT's astronomical performance; it was eventually replaced by a system designed by MSc student Tom Bruce assisted by Tim Chuter in Hawaii.

Telescope users in these early days included both experienced infrared astronomers, some of whom brought their own instruments, and those new to the infrared who used the observatory's common-user facilities. Each of these groups was supported by a staff astronomer who acted as telescope operator, instrument expert and sometimes as a tutor to the visiting groups. The scientific programmes covered a range of topics with the size of the telescope at last making it possible to attempt extragalactic projects, even though getting a K band (2.2 μm) magnitude for a high redshift galaxy might take an hour or more of determined integration. Despite the difficulties of altitude and logistics on such a high mountain site UKIRT gradually began to make its mark as a telescope and establish the UK as a major player in infrared astronomy.

## Stranded

On the 22nd of November 1979, the Thanksgiving Day holiday in the USA, 13 teenagers became stranded in the snow near the summit of Mauna Kea. They were rescued with the help of Kent Tsutsui of UKIRT and Dennis Bly and Adrian Nicol from Cambridge University who took oxygen to the group and helped them back to the road. This incident occurred in fine weather but was the harbinger of a much more dramatic event later in the winter. On the 8th of January 1980, there was a severe storm that caused the road to become impassable, leaving Sidney Arakaki, Peredur Williams and Chris Impey, a Ph.D. student from ROE, stuck at summit. Chris recalls:

'*The stranding story is true and was an intense experience. It was an epic storm that the Weather Service under-estimated, and of course at 14,000 feet we ended up in a gale force driving ice storm. We were preparing to abandon ship as the weather deteriorated and I remember the CFHT Bronco passing us as we were starting ours, but we stalled out, abandoned the vehicle and were stuck*'.

Peredur Williams' diary records that the wind was so strong, and the ice forming so quickly, that Chris and Sidney were blown over as they tried to regain the dome. At the time there was an ambulance stored at the summit near the standby generator hut, just upslope from UKIRT. They managed to get this vehicle started and used the radio-telephone to find out what was happening, first making contact with the fire station in the town of Kona where, by chance, Sidney's cousin was on duty. David Beattie told them that the University of Hawaii observers, who had abandoned the summit at the same time as the CFHT, had found the road very bad, so the UKIRT team were advised to stay put. The power, which was supplied by generators in "Goodrich Pass" lower down the mountain, failed at 4 o'clock on Wednesday morning.

Chris continues, '*There was only rice and beans in a chest, so it was very basic. After power went out, we had to play each other out on ropes into the gale to gather ice to melt for water. We got bloody cold and eventually lit a fire made from wooden pallets in the dome*'.

On the 3rd day, in a howling wind but not actual snow, Peredur Williams and Sidney roped themselves together and worked over the ice to the standby generator hut. It was locked. They "let themselves in" with a crowbar and, after disconnecting all the other telescopes, got a generator started. Power was restored to UKIRT by 5 pm on Thursday. Moving around outside was too dangerous to contemplate due to the winds, but there was a tunnel from the generator hut to the 88 inch telescope building so the pair went across to look for some better food than the rice and beans in the UKIRT cupboards. It turned out that the supplies there were just as unimaginative; more rice and beans. Perry says '*This discovery was one of the more disappointing moments of my life*' and Chris adds '*We concluded that the nobody who put emergency supplies at the telescopes ever thought they would be stuck there. The best item at UKIRT was a set of "space sticks," somewhat indigestible astronaut food put there as an afterthought it seemed*'. The only moment of light humour was the discovery that the 88 inch building had only an electric can opener, which would have been useless had the power gone off.

On both Wednesday and Thursday David Beattie and others had managed to reach the 12,000 ft level but were unable to get any further. At the summit the winds were so high that the snow did not stick, it simply blew straight past, but in the passes between the cinder cones it was drifting and blocking the road. By Friday a bulldozer and sno-cat had been brought into action but even these were still getting stuck on the hairpin bends below the summit. On Saturday they tried again in the morning, returning for another attempt in the afternoon. Alf Neild was in the sno-cat and remembers that it had *'A constant desire to turn left due to a hydraulic leak in the right side control, so we robbed brake fluid from the road grader just to keep going. Not an easy thing to do, in very high winds; and blizzard conditions'.* More brake fluid was brought up from sea level by Susan Beattie, acting once again as an unpaid UKIRT assistant, and for which she was rewarded with a hug and kiss from the bulldozer driver.

**Fig. 6.3** Conditions on the road were appalling as the rescue team tried to reach the summit (Photo Alf Neild)

Peredur's diary records the progress which was reported to the summit by telephone. The cavalry finally came over the hill in mid-afternoon and the party was on its way down about 3 pm. They were home by early evening.

According to Chris '*Dave Beattie led a convoy up the mountain, they had to carve a path in ice that had sloped to form the contour of the cinder cone on the last switchbacks. The bulldozer driver who got to the top was a huge local called Teddi* [Theodor], *who'd never been up high and was badly altitude sick. On the way down, the three of us clung to part of the bulldozer while Dave held on with one hand and with the other held an oxygen mask on Teddi's face, with him carving out the ice, very near the edge and with the road not even visible. It was a wild ride'.*

There was a weather station at the summit that recorded a minimum temperature of −16 °C with a wind-chill factor taking the effective temperature down to −59 °C. The peak gust recorded was 150 mph (240 km/h) but, as the weather station was destroyed in the gale, and the highest wind-speed the station could record was 150 mph; we will never know what the maximum wind-speed actually was. The storm, which of course affected much of the island, caused homes to be evacuated, roads closed and boats sunk across the State. It was such a dramatic episode that, in January 2005, the Hilo Tribune-Herald newspaper included brief details of the astronomers' ordeal on its 'This day in History' items for January 10 and January 12.

**Fig. 6.4** Sidney Arakaki (*front*) and Peredur Williams at Hale Pohaku after their rescue from the UKIRT dome (Photo Alf Neild)

Although The UKIRT team learned from this experience and have ever since had a policy of leaving too early, rather than too late, it was not the only case of close calls with the weather. Bill Parker was working on the summit with Kent Tsutsui when '*We received a warning that the weather was getting worse and the summit should be evacuated. We assumed that all at the summit had been informed and scooted back to Hale Pohaku in worsening weather. The snow at the summit road was already beginning to drift into piles. We got down OK, but then heard that two Japanese astronomers were still in the 88 inch dome. Kent and I considered the situation and reckoned there was enough time to make a trip to the summit before the road became impassable. We scooted up and had to stop on that last acute bend just before the long curving strip of road up to the telescopes, because the snow was drifting too deep beyond that point. We phoned the Japanese guys to check they had a flashlight and said they should walk down the road with all due speed (they would have the wind at their backs) and meet us at the bend. Kent turned the Blazer to face downhill and I started up the hill into the face of the wind to meet the downcomers. I will never forget how the blowing snow particles stung my eyes as I tried to walk into the wind. I couldn't look directly ahead. After I had gone a couple of hundred yards very slowly I saw the flashlight coming towards me. I was amazed that the Japanese guys were dressed in light clothing, no overcoats or gloves. We all piled into the Blazer and after a hairy moment getting past the first snowdrift on the road, proceeded safely to HP.'*

Pat Nelson, Ken Maesato and Charlie Richardson were also briefly stuck at the summit one Saturday after driving up in the morning with a light snowfall being brought in by a strong westerly wind. They kept a careful eye on the weather while working and at about noon they heard from one of the other telescopes that they had decided to evacuate the summit. So they too decided that it was also time to go down as the snow was getting heavier but they then discovered that their Bronco would not start. They had parked it facing downhill into the wind to avoid having to turn around should they need to leave quickly but this proved to have been a serious error. The drifting snow had been blown in through the radiator grill and was packed solid around the engine. When they lifted the bonnet the whole engine space was a solid block of snow with the indentation of the bonnet clearly visible on the snow and no sign of the engine. Soaking electrics and snow in the air intake and carburettor meant that there was no chance of getting the engine going, so after 20 minutes of trying they went back inside, raised the alarm and settled in to wait. All the other telescope crews had either not bothered to come up or had already left, so they were alone on the summit.

After eight hours the sno-cat arrived from Hale Pohaku to pick them up. It was Charlie's only trip in the sno-cat and he says he was glad that it was the only time. The vehicle slithered its way down the summit road through 6 ft (2 m) snow drifts and often felt like it was about to slide off the edge of the road. The passengers also noticed that there were sparks coming from electrical wires on the floor that were right next to the open fuel hose from the petrol tank and this occupied all their attention until they were down off the summit to a point where they could be safely picked up by UKIRT Broncos.

Kevin Krisciunas also learned that discretion was the better part of valour when he had a similar experience. *'One day in April of 1982 I experienced the only whiteout I've ever seen: 15°F, 75 mph winds, and horizontal snow. We abandoned the summit shortly after this extreme weather came up. We did not want to follow in the footsteps of Peredur Williams, Chris Impey, and Sidney Arakaki'*. He also remembers that *'Jim Harwood, who was one of the site testers prior to the establishment of the UH 88 inch telescope, told of a time when he was in a trailer near the summit and they had to get out of the elements for a little while. There was a horrible scraping noise. They wondered if it was a Yeti. Someone sheepishly peeked out, and it was the wind peeling the paint right off the side of the trailer'*.

## The Ford Popsicle

One day when Kent Tsutsui and Alf Neild were at the summit, a blizzard caught all the day crews by surprise and threatened to snow everyone in. The other telescopes were calling UKIRT to check on the road conditions and IRTF was asking if the UKIRT crew could see one of their vehicles that had left to check the road conditions and had not returned. Alf went to look but by then it was snowing so hard that he could not see very far. At about the same time, the CFHT crew was finding out that their part of the road had become so bad that only two of their vehicles were able to make it to the crest. Deciding to walk everyone else out, they called UKIRT to meet them at the UH 88 inch telescope since the University of Hawaii day crew had already left. As there was also the IRTF staff to contend with, Alf and Kent split up, with Kent taking one of the Ford Broncos to assist CFHT while Alf went to help with IRTF. Alf decided to drive his vehicle down the IRTF road and meet the crew as their lost vehicle had become stuck somewhere in the middle. In this attempt, his Bronco became stuck in the snow and an attempt to reach it with the second UKIRT vehicle was abandoned when it became clear that it too would get stuck.

Eventually both the IRTF and CFHT crews reached the crest of the hill by the 88 inch telescope and after accounting for everyone, the three telescope crews squeezed into the three remaining vehicles and evacuated the summit in a continuing heavy snowfall. Alf and Kent stayed at Hale Pohaku as they were on shift that week. Then David Beattie came up and "recruited" them to go back up to the summit as soon as there was a break in the weather in order to refill the instruments with liquid nitrogen. The weather cleared at about 2030 hours and the three men headed up to the summit. They managed to reach UKIRT where David Beattie headed to the dome and Alf and Kent went to the stranded Bronco on the IRTF road, now hood deep in snow. With some effort they managed to get the Bronco just about running. With a rough running engine, they managed to drive it to the concrete parking pad at UKIRT, where it died again. As the weather started getting bad again, the three men then abandoned both the summit and the Bronco.

When the weather finally cleared and it was possible to get back to the summit, the abandoned Bronco was found with a thick build up of ice on its side. After some work, Kent and others were able to drive the Bronco out, leaving a block of ice with the Bronco's imprint in it. The rest of the day was spent cleaning up inside the dome and when locking up for the day, it was noticed that the ice block had disappeared and was being hauled away in a pickup truck that was leaving the summit.

**Fig. 6.5** Pictures of what became known as the "Ford Popsicle" taken after a storm in February 1982 (Photo Alf Neild)

## Science Starts to Flow

The next major step forward was the commissioning of the f/35 chopping secondary mirror in the summer of 1980. The f/35 secondary was mounted on a separate top-end, but the limited space in the small dome meant that swapping the top-ends over was not easy. To make the swap it was necessary to position and tie down the telescope at its southern limit before detaching the top-end and hauling it up into the dome roof with hand cranked winches. Great care was needed because the clearances were very small and, according to Terry Lee, if a cable snapped it might decapitate the person cranking the winch (for this reason the winches were motorised later). Once at the top of its travel the dome was rotated until the unit was clear of the telescope and could be lowered into its parking spot on the dome floor. The replacement could then be picked up and the reverse procedure carried out to install the new top-end. It was an operation requiring four or five people and taking half a day to complete.

Other infrastructure improvements included the installation of a helium liquefier in Hilo. This new facility, which could supply liquid helium for any of the telescopes on the mountain, was commissioned by Sidney Arakaki and according to the 1980 ROE annual report greatly reduced the *'uncertainty and stress previously associated with the procurement of liquid helium'*. Soon to follow was a new telescope control program written in Fortran, rather than Assembler, and which promised to be much easier to develop. The new software offered additional capabilities including the ability to precess star co-ordinates, to spiral-scan the telescope and to slave the dome to the telescope so that the dome followed the telescope as it tracked.

That results were starting to flow was evidenced by papers presented to the IAU symposium No. 98 on Infrared Astronomy, which was held in Hawaii in June 1980. The first refereed UKIRT papers were also being published, perhaps the first being *Bailey, Hough and Axon, IR Photometry and Polarimetry of 2A0311-227* in Nature on the 29th of May 1980. This was soon followed by the first paper to be published in the Monthly Notices of the Royal Astronomical Society, it being *Near Infrared Spectrometry of WC stars by Williams, Adams. Arakaki, Beattie, Born, Lee, Robertson and Stewart. (MNRAS (1980) 192 25-30P)*. Another milestone was the first UKIRT symposium, held on the 15th and 16th of July 1981 in the David Hume tower of the University of Edinburgh. Set up to emulate the better known AAT symposia and establish a forum for discussion of UKIRT and its results, it was attended by over 80 people, including some of the UKIRT staff from Hawaii. A total of 28 papers were presented.

In January 1981, with the telescope in operation rather than under construction, the SRC replaced the UKIRT Steering Committee by a UKIRT Users Committee with Ian Robson, then at Preston Polytechnic, as Chairman. About the same time an article called "Looking for hot spots in deep space" written by ROE observatory secretary Bennet McInnes appeared in Issue 166 of the British Council magazine SPECTRUM (previously known as British Science News). In New Zealand a young Stuart Ryder saved a copy from a grisly end. Stuart writes *' It somehow found its way to the ends of the empire to Kings High School in Dunedin, then must have been destined for the rubbish bin (I didn't steal it from our library, honest sir!)'*. Twenty years later, Stuart would join the UKIRT staff. Not long after, Ian Robson visited the Royal Aeronautical Society's branch at Warton airfield in Lancashire to give a lecture on infrared astronomy. Flight test engineer John Davies was in the audience and was rather taken by the idea of working at a telescope up a mountain in Hawaii, setting up a strand that would cross many times in the years ahead.

**Fig. 6.6** Sidney Arakaki topping up a dewar with liquid nitrogen (Photo Peter Forster)

Over time the UKIRT instrumentation continued to improve. Leicester University astronomers were involved in the construction of a visible (UBVRI) photometer, which would be commissioned at UKIRT in 1981. Other instruments were provided by the UK sub-millimetre community, based mainly at Queen Mary College, London (QMC) but also at the Mullard Radio Astronomy Observatory (MRAO), Cambridge. These groups were keen to use UKIRT as they had found that other telescopes were at too low an altitude or lacked a large enough collecting area for successful sub-millimetre observations. UKIRT with its high altitude site and large collecting area, promised to be an ideal telescope with which to open up this emerging field.

The astronomers at London had pioneered continuum sub-millimetre astronomy using facilities at the Pic du Midi in the French Pyrenees, Mount Evans in Colorado, Kitt Peak in Arizona and at the UH 88 inch telescope on Mauna Kea. In January 1980, they used UKIRT for the first time with the QMC photometer, which had originally been developed for use on the UH 88 inch telescope. Since the QMC photometer had a large beam, the tracking problems then affecting UKIRT were not too damaging and the overall performance of the instrument was satisfactory. The observing team comprised Ian Robson (who by then had moved to Preston) and Peter Ade. It was the first of a large number of visits, which culminated in the arrival of the common-user sub-millimetre instrument UKT14 in 1986.

Another group of sub-millimetre astronomers, using a heterodyne receiver jointly developed by QMC and the MRAO, were also among the users of UKIRT during the winter of 1980/81. The observers were Glenn White, Peter Philips and Graeme Watt from QMC and Richard Hills and student Russell (later Rachael) Padman from MRAO. This activity was the beginning of a long series of heterodyne sub-millimetre astronomy programmes at UKIRT, with the QMC/MRAO groups being later joined by a team from the University of Kent led by Les Little. These teams sometimes did their observing during the day, which is possible because at sub-millimetre wavelengths the sky is equally bright at all times, day or night. However, although daytime observing at these longer wavelengths is theoretically possible; it is not that easy in practice. Nonetheless, these groups were anxious to use as much of the telescope time as possible and sub-mm work continued until a dedicated millimetre telescope took over this role in 1987. Thus UKIRT brought huge versatility to the UK community, not only undertaking frontline astronomy from 1 mm to the optical but also providing a training ground for astronomers and their students. The result was a steady improvement in instrument capability, and hence scientific progress, especially in the UK.

Another pair of important visitors was Dave Aitken and Patrick Roche who brought their 10 μm cooled grating spectrometer to UKIRT which, at the time, lacked its own such instrument. Aitken and Roche's instrument was a success and would eventually lead to a common-user 10 and 20 μm spectrometer being developed for UKIRT, but the same could not be said of the Michelson Fourier Transform Spectrometer (FTS) instigated by Jim Ring and brought to UKIRT in 1981 by Richard Wayte and Helen Walker of Imperial College of Science and Technology, London.

Wayte's instrument was designed for high resolution spectroscopy in the visible region and had its origins in a prototype which he and his student Ian Wynne-Jones built and installed on the 2.5 m (98 inch) Isaac Newton Telescope in Sussex. With this instrument they discovered the hyperfine splitting of the interstellar Sodium I D line, but the telescope was really not big enough, nor the British weather good enough, for the instrument to succeed. So approval was given for Wayte to build a Michelson FTS for UKIRT. The instrument was built on a steel optical bench about 3 m long and weighing ¾ of a tonne, so shipping it to Hawaii proved to be an interesting challenge. The type of cargo plane with doors wide enough to get the crate on board was too large to land at Hilo airport and so the package made the last part of the journey by ship (on the top of the deck judging by the corrosion to the stainless steel discovered when it arrived). The crate was eventually taken to the summit and was installed in the Coudé room at UKIRT. With help from several of the UKIRT engineers the Imperial College team unpacked, repaired, cleaned, repainted, re-assembled and then tested their instrument using light fed from the secondary focus to the instrument via an optical fibre. During commissioning the weather was poor and the stars observed happened to be ones where hyperfine splitting was not observed, so the project was terminated when the commissioning run was finished. The instrument languished in the Coudé room for several years and the crate became a cable store (earmarked as emergency firewood, if necessary). Helen Walker remarks that '*It must be one of the most unlucky instruments of all time. By the time it was built and shipped out to Hawaii, the composition of the UKIRT committees had changed, and they no longer wanted an optical instrument on an infrared telescope*'.

## Altitude Effects; "Sickness and Thickness"

Once the telescope construction on Mauna Kea starting to approach its conclusion, the issue of understanding any effects of altitude on UKIRT staff and visiting observers became more pressing. The main concern of the SRC was, of course, to safeguard its personnel, but it was appreciated that if, in so doing, useful medical data on the effects of altitude could be obtained that would be a considerable bonus. The issue was a tricky one, the risks were not well understood but at the same time there was a feeling in some quarters that the UK was making more of a fuss about them than other Mauna Kea users since the University of Hawaii's 88 inch telescope had been operating on the summit since 1970. Discussions were arranged with Professor Donald Heath of the University of Liverpool, a recognised authority on high altitude pathology, to see if a mutually acceptable plan could be developed. Heath proposed a dedicated study over two years to be carried out by a physician from the UK under his supervision. However, SRC management felt that the plan he proposed did not require a dedicated appointment and that the tests required could be carried out as effectively and more cheaply by a local medical researcher under the joint supervision of Heath and Professor Irwin Schatz of the John A. Burns School of Medicine in Honolulu. At first it seemed that this arrangement would be satisfactory, but it turned out to be more complicated than initially thought. Vincent Reddish argued for an ROE appointment to assure that the local staff felt the medical researcher was indeed a part of the UKIRT team and was conducting a study of mutual interest and value. There were also concerns about the implications on the other groups, on the attitude of the Hawaii staff who felt they were able to look after their own affairs without an overprotective management at ROE telling them how to behave and on the relationship between Heath and Schatz over who would really control the project.

In the end it was decided to appoint a UK based researcher who would be posted to Hawaii for 2 years and that the appointment would be made by Liverpool University under contract to SRC. So, in June 1980, Dr. Peter Forster, a medical research fellow in the Department of Pathology of the University of Liverpool joined the project team in Hawaii. Forster's study was focussed on the physiological effects of working at altitude. At its core was a comparison between UKIRT staff, who worked a shift system based on several five day stays at Hale Pohaku followed by a long break at sea level, with IRTF staff members who commuted up the mountain on daily basis. The astronomers were a unique population for such a study being, in the main, unadapted to living at altitude and unlike mountaineers and the like, oscillating regularly and in a short time between 0, 9000 and 14,000 ft. (0, 2800 m and 4300 m).

The UKIRT shift arrangements were based on a plan formulated by Terry Lee in 1979 who wanted *'To have at all times on the mountain personnel to cover mechanical, electronic, software and instrument support skills as well as a staff astronomer with the observing team'*. The shift system he devised consisted of 40 days at sea level interspersed with 40 days of mountain work. The mountain work comprised four cycles of 5 days of summit work preceded by a night of acclimatisation and followed by a 5 day rest period at sea level. There were several shift teams, each with an astronomer a software engineer, a mechanical technician and an electronics technician. The technical team would mostly work at the telescope during the day, with the astronomer and perhaps some other staff members going to the summit at night. On the last day of each shift a new team would come up to acclimatise and there would be a brief overlap before the old team went down to sea-level.

Forster's research programme included questionnaire based assessments of how people reacted to altitude and quantitative measurements such as weight, blood pressure, respiration rate, ECG tests, urine and blood samples and simple exercise tests (e.g. stepping on and off a box 30 times in a minute). There were also series of psychometric tests, which included being read a series of numbers and then quoting them back from memory, first in order, then in reverse order. Alan Pickup scored higher at altitude than at sea level which he ascribes to familiarity with the test, which he first did in Hilo, rather than some bizarre physiological response to altitude. Some of the astronomers managed to get up to around a dozen numbers in reverse, causing some to believe that that there was a private competition going on to see who was best at this exercise. Andy Longmore remembers Forster as a very considerate man who, when blood was required would ask "Which arm would you like me to use", a question which once provoked Malcolm Stewart to point at a co-worker and say "His"!

Some of the medical equipment had to be repaired or adapted to high altitude work by the UKIRT technical staff. Forster had a blood gas analyser on loan from the School of Medicine in Honolulu in which the digital display failed. Bill Parker had a look at it and brought out a couple of wires so that a digital multimeter could be used for the display. It was a simple thing for an electronics technician to do, but Bill got the impression that Peter seemed to think that he deserved a Nobel Prize for this piece of work. Bill and mechanical engineer Rory Urquhart also made some test gear so that Forster could check the calibration of a gas velocity measuring device, which he used to measure expiratory flow rates at sea level and at altitude. They built a small pump out of a plastic bag and a piston with some very simple circuitry to measure the piston velocity and coupled it to an oscilloscope to measure the velocity of the piston. The study was published as a letter to The Lancet with Bill Parker as co-author.

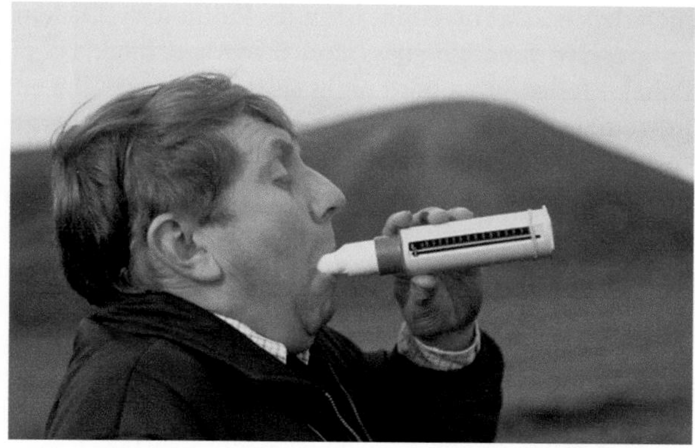

**Fig. 6.7** Bill Parker's lung function is tested with a peak flow meter (Photo Peter Forster)

Peter Forster published his work on the medical effects of chronic intermittent exposure to high altitude in a series of papers in medical journals and in a 1984 paper (PASP 96 p 478–487) for a non-medical readership. A full description also exists as Occasional Reports of the ROE number 11, a 98 page typescript with 57 tables and 23 figures. This report ends with the remark that *'It has been a pleasure to witness the UKIRT staff in the pursuit of excellence'* and includes a series of recommendations which would become very familiar to regular UKIRT users. For example it recommends the need for, and frequency of, high altitude medicals, a list of "Red Alert" symptoms for altitude sickness, a rule that no-one should work alone at the summit and another that a vehicle should always be available for evacuation in the event of the early warning signs of altitude sickness. Many of these, including a ban on visitors under 16 years of age, were taken up and have remained in force ever since.

Another conclusion was that the shift system of technical staff spending five or more nights at mid-level and commuting to the summit offered better efficiency than the much longer sea-level to summit commute favoured by the IRTF staff. Indeed Forster recommended that shifts of more than the 5 day cycle would be better physiologically, although he conceded that the enforced separation from family at sea-level was a strong argument against longer captive shifts at Hale Pohaku. The shift system would operate for a number of years before being abolished, apparently as a cost saving measure, and replaced by a weekday commute of the UKIRT day crew by the mid-1980s. This move, which originated from administrators in the UK, eliminated the allowance that the shift workers had received to compensate for time away from families and for working such irregular hours. Instead staff had to claim for the actual hours worked. By several accounts this caused considerable resentment and, ironically, probably cost more money because of the prodigious amount of overtime that was being worked.

Forster's detailed report listed a dozen or so specific medical incidents affecting (anonymous) UKIRT and IRTF staff which had occurred on Mauna Kea. They included a case of High Altitude Pulmonary Oedema. This manifested itself as progressive breathlessness and a marked blueness of lips that occurred at Hale Pohaku following a third night shift at summit. The emergency was resolved by an immediate descent and the administration of oxygen. Other staff and visitors suffered migraine headaches and bronchopneumonia. These case reports illustrate that despite all the precautions and operating guidelines, Mauna Kea was not a benign working environment and indeed high altitude exposure did take the life of one astronomer at another facility: a 37 year old smoker and diabetic suffered a heart attack after staying at the summit all night and working through the days. Post mortem examination revealed extensive coronary artery disease and triggered a programme of regular health checks for the UKIRT personnel that continued long after Forster left.

At a more minor level most people suffered from headaches, breathlessness, insomnia, lethargy, poor concentration and impairment of memory, but these symptoms usually passed after a day or two at altitude. Not everyone was so lucky. One person found it impossible to keep food down if he ate just before ascending to the summit or ate once he got there, and for some there was no solution. One staff member became sick on the first day of work at summit on every shift, necessitating a descent to lower altitude. Attempts to minimise the symptoms by gradual staging of ascent to summit were without success and eventually this person was excused from mountain duties entirely.

In addition to the effects of altitude, concern was also expressed about inhalation of the volcanic dust thrown up when driving on the unpaved road between Hale Pohaku and the summit. The worry was the dust might contain fibres similar in composition and form to the thin, needle shaped (amphibole) fibres of asbestos and, if so, whether inhaling fibres would pose a health risk. Samples of dust were collected at different roadside locations and sent back to the Pneumoconiosis Unit in Wales for testing, but no suspicious fibres were detected. Nevertheless a few individuals took precautions: some insisted in driving on the summit road with the car windows shut and one individual wore a face mask, although the effectiveness of this measure was brought into question because he cut a hole in the facemask to allow the insertion of a cigarette!

There were other hazards too. Dave Robertson got snow blinded and snow burned one day. He says, *'We had to go from about 11,000 to 14,000 feet in about 3 feet of snow. We met a bank of snow [on the road] so we walked up the rest of the way. I had no sunglasses and my hood was only covering part of my face. They could peel the skin off my face when we got down, my eyes got scarred and I had fairly bad eyesight for a couple of weeks'*. When asked why they had even attempted to walk up he said simply. *'It seemed like the thing to do at the time. It was that or be stuck at Hale Pohaku'*.

What would never be written up, until now, was some of the less quantifiable effects of altitude. Here are some recollections from members of the UKIRT team. Perhaps fortunately, the names of those involved have mostly been lost to history.

Alf Neild remembers taking a "Newly Arrived Astronomer" up to the summit so that he could get his detector ready for that evening's observations. When they arrived at the summit they went to the common room for a 10 minute rest. The visitor wanted a drink of coffee but the first jar of coffee was empty, so he grabbed a full jar and screwed the top off it. He then picked up a spoon, to stab it through the jar's top seal. When he stabbed the seal, there was a mini explosion of coffee granules; which found their way up his nose, and in his mouth and hair. What he had failed to realise was that at sea level coffee jars are under vacuum, but in the thin air of the summit they are under pressure.

One day a visitor taking photos inside the UKIRT dome ran out of film. He changed the film, and carried on taking photos before returning to sea level. Two days later, he called to report that he had a problem with his camera. Apparently, it was an airtight underwater camera; so when he changed the film and closed the camera back, it had the partial vacuum of Mauna Kea inside it. When he returned to sea level, the camera was under vacuum and he could not release the camera back. He had to return to Hale Pohaku to open it.

Alf Neild met another "Newly arrived at the summit astronomer" in the common room, and asked him if he felt OK at altitude. The visitor said "Altitude does not affect me at all, as I normally run 5 miles a day" and then proceeded to fill the electric kettle and switch it on. He then picked up the kettle, placed it on the oven hot plate; and switched the hot plate on as well. The first Alf knew of this mistake was the acrid smell of the kettle's plastic feet melting.

Another visitor arrived a few weeks earlier than the start of his allocated observation time in order to set up and calibrate his instrument. He complained that the noise and vibrations from three vacuum pumps, located in the north column of the telescope, were upsetting his calibrations and he wanted them switched off. As the pumps were being used to pump the helium cryostats of the instruments in use, this was not possible. After much discussion he was told that the pumps must stay on until his own observing run started. Unbeknown to the staff, and unaware of just how hot such pumps can get in such the rarefied atmosphere of the summit, the visitor placed pieces of polystyrene packing sheets under the pumps to try and stop the vibrations. Some time later, the dome filled with acrid smoke, and there were flames coming out of the north column. As there were a lot of electrics in there, a powder extinguisher was used to put out the flames which, incidentally, also contaminated the visitor's instrument.

One ROE staff member used to tell this story about his first trip to the summit. At one point he wanted a drink of water, and saw several cups with people's names on the back of the toilet in the downstairs bathroom. He decided to use one and poured himself a drink, and only later learned that these were cups for collecting urine specimens from the staff!

Muddled thinking was not however exclusively the realm of astronomers, anyone could join in. One night Peredur Williams and Bill Zealey were observing when the computers died. UKIRT had a contract for on-call computer support and since it was still before midnight they called out the computer technician who arrived about 2 am and started to take the control computer to pieces. Having nothing to do Bill and Peredur went for a coffee and returned to find the technician surrounded by computer boards and looking sorry for himself. Having taken the machine apart, he could not remember how to put it back together.

**Fig. 6.8** Unnoticed by most people, even when asked to study the sign carefully, the "T" and "I" of this summit warning sign are reversed (Photo Ian Robson)

However not all of these incidents were from visitors. Neild himself recalls '*Just after the storm of 80/81 the dome ventilation units became clogged with large ice formations, which had to be removed. I togged up with all the gear, and spent the next 5 hours on the dome roof; slowly chipping the ice off the vents. It was only when I got down that I noticed my left ear had been outside my cap, had got second degree sunburn, and looked like a braised lamb chop*'.

On another occasion Terry Lee specified that a 4 inch diameter line be installed from the south column to the mirror cell for pumping on liquid helium. When he arrived on the summit to install the bolometer detector cryostat he was surprised to find four 1 inch lines plumbed from the pump to the mirror cell! Seemingly Alf had failed to find the 4 inch tubing and used his initiative.

On one of Charlie Richardson's first visits to the summit he decided that he wanted a snack. He found a small tin of beans in the cupboard and decided to heat this up in the microwave oven. Thinking he was being sensible he made a hole in the top of the tin first, then put the tin on a plate and put them into the microwave to heat up. About 20 seconds later there was a load bang as the plate shattered. Switching off the microwave he then grabbed the can which was red hot. Looking back on this he ruminates that *'I have a degree in physics, although you would not believe it from this incident'*.

Dave Robertson recalls another, typical, incident of muddled thinking at the summit which occurred when cleaning the inside of the dome for the dedication ceremony. *'A group of local cleaners came up to help with this. We wanted to get all the dust off the inside of the dome and while we had some long poles they were not long enough. So we found a pallet truck and we put a couple of chairs stacked on top of that and one of the locals climbed up, balancing on this. We started pushing him around so he could clean, then we pushed him around a bit more and so on. When we reached the vestibule with its flat roof covering the control room and the entry door it was possible to climb up onto the vestibule, so you didn't need this contraption to stand on [to reach the dome]. Now of course the dome* ***rotates*** *so we didn't have to push this guy around, but collectively we all came to the conclusion that we had to put him on top of this thing and push him around'*. They had all failed to realise that someone could simply get up on the vestibule roof and clean the dome as it was rotated past.

Ian Sheffield recounts this incident which happened a few years later. *'For ISU2 we needed a set of cables going between the control room and the telescope, so we asked the local staff if they could drill out a 2 inch hole between the control room and the telescope and they said "Yes, we'll do that". So we went up and spent ages looking for this hole, but we couldn't find one. So we got a drill and drilled one ourselves. Half an hour later, literally 2 feet away from the hole we had drilled, we found the hole which they had made'*.

Another incident occurred some years later, although the precise details are remembered slightly differently by the people who were there. Telescope operator Dolores Walther recounts that *'We had a person who wasn't feeling well and was sitting in the staff lounge. He was looking a bit green so the safety officer intended on administering oxygen, but what he picked up was a fire extinguisher that he tried to administer to the patient'*. Andy Longmore, the safety officer in question, has a slightly different account. He says that he noticed that a visitor who he had brought up, and who was not fully acclimatised, seemed to have gone missing. He set out on a search and, intending to take an oxygen cylinder "just in case", did indeed pick up a fire extinguisher by mistake.

## Co-operation with the Netherlands

In the late 1970s and early 1980s UK and Dutch astronomers were involved in a number of collaborative projects. The Northern Hemisphere Observatory had at last found a home in La Palma and was in the process of being set up with the Netherlands taking a 20 % share in its three telescopes. At the same time Dutch astronomers were proposing an infrared sky survey mission that would evolve into the US/Netherlands/UK Infrared Astronomical Satellite (IRAS). Since the Dutch were investing heavily in IRAS and the UK had deployed considerable efforts on UKIRT, it was logical that SRC and the Dutch community might benefit from closer collaboration. Encouraged by informal contacts with Terry Lee, Reinder van Duinen from Groningen University wrote to Harry Atkinson at SRC in 1977 to start discussions about Dutch access to UKIRT to follow up results from IRAS. Following on from this approach van Duinen met with Terry in November of 1977 to investigate possible collaborations between ROE and Groningen related to multi-element photometers which might be mounted at UKIRT for following up sources detected by IRAS once it was in orbit. UK scientists, notably Dick Jennings from UCL and Peter Clegg of Queen Mary College had a strong interest in this kind of project and a meeting of an SRC working group on far IR and millimetre wave astronomy endorsed UK participation in IRAS.

Eventually arrangements were put in place for the Netherlands to contribute to UKIRT operations in return for observing time during the period up to end of the IRAS mission. The Netherlands community would contribute the equivalent of 15 staff years of effort towards the cost of UKIRT infrastructure and operations. The staff contribution would be delivered by the secondments of Rich Isaacman (Leiden) from the beginning of 1981 to the end of 1983 and Tom Geballe (Groningen) between September 1981 and August 1984. Tom admits to knowing '*Next to nothing*' about UKIRT when he accepted the offer to move to Hawaii in 1981. His decision was '*Born partly out of adventure, partly out of previous pleasant experiences on the Big Island and partly out of desperation*' since his third post-doctoral fellowship was drawing to a close and he had no other tangible job offers. It is a little ironic that the first two of these "Dutch" astronomers were in fact both expatriate Americans working in Holland. Tom Geballe did not realize this at first; Rich Isaacman was fluent in Dutch and after hearing him on the phone talking to Leiden, Tom concluded that he was a native Dutchman who spoke excellent English rather than the other way around. In addition to these two, John O'Sullivan from Dwingeloo would provide consultancy support regarding telescope performance. Additional contributions were to come in the form of a new chopping secondary drive, a common-user sub-millimetre wave receiver developed by Thijs de Graauw of the Dutch SRON laboratory plus the staff effort required in Hawaii to commission the instrument. In return the Dutch community would be allocated a seat on the UKIRT panel for allocating telescope time with the expectation that about 15 % of the time would pass to Dutch astronomers.

**Fig. 6.9** Tom Geballe in the Leilani St office (Photo ROE)

The decision in 1981 for the Netherlands to join in plans to build a millimetre telescope (which would become the JCMT) on Mauna Kea in collaboration with the UK (and eventually Canada) gave impetus to extending this arrangement past its rescheduled end date of 30 September 1984 (the original end date had been the 31st of December 1983 but delays to IRAS and the phasing of the Dutch observing time had pushed things back by a few months). The new protocol would be similar, with the Netherlands contributing the equivalent of 6 staff years of effort and some further hardware in return for a share of about 5 % of the observing time for a further 3 years. The staff effort was to be provided by a further 3 year tour by Tom Geballe plus the appointment of a Netherlands funded research fellow to the Hawaii site. The sub-millimetre instrument would be transferred to the JCMT to become one of its first light instruments.

Although it takes us a little ahead of our story, the arrangements were extended again to cover the period after 1987 with a different medium of exchange. This later agreement was for the trading of Dutch time on the JCMT and the La Palma telescopes for nights at UKIRT according to a formula determining the relative value of the respective nights. These values were set as UKIRT = 1.0, JCMT Shift (8 hours) = 1.4, WHT = 1.3, INT = 0.5, JKT = 0.15 with appropriate weighting by PI and Co-I in the case of joint proposals.

## Computing Improvements

As UKIRT evolved there was steady development of both the computing hardware and software so the number of software staff slowly increased. Kevin Krisciunas, who in 1980 was writing software for NASA's Kuiper Airborne Observatory (KAO), was asked by Ian Gatley if he was interested in coming to work in Hawaii. He joined the UKIRT software group in March 1982 but his first reaction was not favourable, writing in 2011 he says that *'My professional opinion in 1982 and 1983 was that the data acquisition code we used at UKIRT was not supportable. It was a real rat's nest'*. He added that when he arrived he asked *'How often do you guys back up the system?'* and they replied *'"Back up? What's a back up? We have it on disk"'. I felt I had taken a big step backwards compared to the KAO programming team. However, once 1985 rolled around, we had professional coding standards and supportable, modular code'*.

Ian Gatley had conceived a system of lookup tables that would allow UKIRT to take a measurement with a given instrument and know right away if the throughput of the sky, telescope and instrument was as expected. As well as giving an indication of the sky conditions, it became possible to use similar software to take a whole night's worth of measurements and automatically derive the atmospheric extinction and zeropoints (measurements of the sky's transmission on that night). With this information the software, which was called PHOTOM, could produce calibrated and scientifically useful photometry by 10 o'clock the following morning, ready for the astronomers when they woke up around lunchtime. According to Kevin, *'Bill Zealey once said that PHOTOM was the greatest astronomical data acquisition software he had ever seen. I thought Bill needed to expand his horizons. From the programmer's standpoint PHOTOM had too many Band-Aids on top of Band-Aids'*. Terry Lee was however more flattering of this package saying that while it *'Might have looked pretty ugly from a software engineering point of view it did most of the things you might want to.... As well as controlling [various instruments] it also issued commands to the telescope, controlled the guider crosshead and later could send optimised waveforms to the telescope to reduce nod times'*.

UKIRT's PDP computers were eventually phased out in favour of Digital Equipment Corporation (DEC) VAX computers which allowed software to be written and tested in modules, which was a huge improvement on the code used by the PDPs. The original VAX 11/730 computer had a seemingly gigantic 120 MB internal hard disk drive. Rich Isaacman remembers Charlie Richardson proudly announcing that that drive capacity was so huge that *'We'll never fill it up!'*. The disk was not however supplied by DEC, who would not certify that their disks would operate in the thin summit air. Peredur Williams borrowed a disk drive from System Industries for testing in the UKIRT building and he recalls simulating a power cut by simply pulling out the plug. The drive survived this rather brutal treatment and so was adopted for summit use.

**Fig. 6.10** The early control room layout was very cramped. Ian Robson and Peter Ade are seen observing in April 1981. The headsets allow communication with a student in the dome (Photo Ian Robson)

## Changes at Home

The arrival of Malcolm Longair as the new Director of ROE in late 1980 resulted in a number of changes, and one of those was to be in the management of UKIRT. When Malcolm took over Ray Wolstencroft had been in charge of the UKIRT Unit, also known as the UKIRT "Home End" since 1976, but as noted earlier relations with Hawaii had not always been smooth. Longair soon changed the reporting structure so that both the Astronomer In Charge in Hawaii and the Head of the UKIRT Home End reported directly to him. Although Terry Lee had been expected to return to the UK before Christmas, over the summer of 1981 Malcolm Longair reconsidered the plans put in train by Vincent Reddish for Ray to take over from Terry. Malcolm decided that Terry would remain in Hawaii and instead Ray was offered a chance to go to the IfA in Honolulu for a sabbatical year, the intention being that Ray would gain more experience with infrared techniques before returning to ROE and continuing as the head of the UKIRT Home End.

Accordingly Ray ceased to be head of the UKIRT Home End in June 1981 and, along with his family he departed to Honolulu soon after. He was temporarily replaced by Tim Hawarden, a South African astronomer who had been working at the UK Schmidt Unit in Australia before coming to ROE. With the structural changes made by Malcolm Longair, Tim would have a liaison, rather than management, role but it was one in which he soon established himself. Years later Terry wrote '*Tim and I got on very well and I had absolute confidence in him both in letting me know what was happening at ROE and in the UK and in setting about whatever we agreed. Tim became an organic part of UKIRT and he embraced the job. He communicated freely with directly relevant committee chairs, they knew what his role was and they saw the results. Tim was one of the great pillars of UKIRT, extraordinary people who made great contributions, who lived the dream'*. About the same time, Frances Shaw was transferred from the UKIRT unit to the central ROE administration group, a move perhaps related to fact that she and Tim were becoming "an item" and, with Tim as head of the group and so her manager, there was a potential conflict of interest.

During his sabbatical in Honolulu Ray was visited by Malcolm Longair who asked him if he wanted to continue with his UKIRT Home End responsibilities when he returned to ROE. By then Ray had become disenchanted with the arrangements and declined, so he came back to Edinburgh in the late summer of 1982 as head of a newly created Research Co-ordination division. Tim Hawarden, who eventually married Frances Shaw, continued as head of the UKIRT Home End in Edinburgh for the next 6 years until he moved to Hilo in late 1987. Richard Wade took over the UKIRT Home End at ROE in August, just ahead of Tim's departure.

## Sky at Night

In December 1981 Patrick Moore and the BBC "Sky at Night" team paid a visit to UKIRT and recorded interviews with Alan Pickup and Terry Lee. As an amateur astronomer and former president of the Astronomical Society of Edinburgh Alan already knew Patrick Moore, but as far as he can remember there was no specific plan for him to be involved in the programme. He says *'Patrick was up at the summit filming when I was on shift and he spoke to me. It was just a co-incidence I think. I had no idea he was going to speak to me until it happened; it was all done on the spur of the moment'*. Alan was filmed standing under the telescope talking about how it worked and Terry spoke about his scientific interests while seated in the telescope control room. A clip of Alan talking was broadcast on 25 April 1982 as part of a Sky at Night 25th anniversary special programme and re-appeared with the others on the 6th of March 1983 in an episode called "Half way to Space". This episode also featured interviews with Dale Cruikshank at the IRTF, a visit to CFHT and a chat in the open air, just downhill from UKIRT, with John Jefferies. Alan Pickup didn't see the finished 1983 programme until 2010 when a copy turned up during the writing of this history.

# Chapter 7
# Consolidation

## Encoders and Instruments

In May 1982 the telescope was unavailable for two days while new encoders were fitted. The new encoders, with 22 instead of 20 bits, improved the tracking considerably but they were not very reliable. They used a large number of filament lamps to illuminate the moiré fringe system inside them and this proved to be a problem because the lamp life was not long and just one failed lamp would render the encoder inoperable. There was a complete set of spare lamps built into the encoder so that if one lamp failed it was possible to switch over to the second set and keep working. However, every time a lamp failed it had to be replaced immediately after the night's observing and this happened every few weeks. Eventually John Clark, Alf Neild and the maintenance team could swap them in a few hours.

By now the telescope was being operated in f/35 mode most of the time and UKT5 and UKT6 had become the standard InSb cryostats. The first Cooled Grating Spectrometer (CGS1) was commissioned by Richard Wade, Tom Geballe, Tim Hawarden and Malcolm Smith in early 1982 adding a new capability to the instrument suite. Although only single detector instrument, it was able to map shocked molecular hydrogen in the Galactic Core by scanning the region of interest. This scanning required considerable precision in telescope pointing so a technique was evolved in which the autoguider was locked onto a guide star visible in the TV camera and then, instead of commanding the telescope to move and relying on its position encoders, the crosshead which held the TV camera was slowly moved. This forced the telescope to follow the guide star and to be moved across the sky with far more precision than could have been done using the encoders themselves. CGS1 was later used to make similar maps of the star formation region in M17, the Omega Nebula. CGS1 was however not long for the world, it was replaced by a new spectrometer called CGS2, which had a seven element detector array, in January 1983.



J.K. Davies, *The Life Story of an Infrared Telescope*, Springer Praxis Books,
DOI 10.1007/978-3-319-23579-0_7

Many of UKIRT's instruments would be built at ROE, which was, and would remain for more than a decade, the home base of the UKIRT operation until full responsibility for UKIRT was transferred to Hawaii. The Edinburgh site quickly acquired an excellent reputation for building infrared instrumentation, a result which many attribute to the efforts of its Director, Malcolm Longair. Malcolm says that when he was approached about becoming Director, the SRC chairman Sir Geoffrey Allen told him that his task was to get the new science out of the ROE facilities. His approach was to completely change the direction of the ROE instrumentation programme to concentrate on a small number of innovative programmes such as the UKT14 bolometer and, later, the first infrared array instruments.

There were two other key players in this transition who were already in place when Longair arrived. Malcolm Smith had been appointed Head of Technology by Vincent Reddish in 1978 but said that he would only take the position on condition that ROE also employed a really good engineer to assist him. This was agreed so Malcolm said that he would start his new job on the same day as this new engineer arrived. It took almost a year to make an appointment and so Malcolm stayed in his job at the Anglo-Australian Observatory rather longer than he expected, finally coming to Edinburgh in 1979. The engineer who was hired was Donald Pettie and this turned out to be a crucial decision because Donald had experience of project management which was more formal than anyone then at ROE. Malcolm Smith himself admits that '*Most of my experience had been with a scientist and 3 or 4 other people, so perhaps a team of 5 or so but with no real project management. Donald knew how to manage projects and how to deliver them. Things like safety and the boring bits, how to pack an instrument before we ship it*'.

Between them the two Malcolms and Donald Pettie stopped the many other ROE instrument programmes which risked interfering with Longair's intention to concentrate on doing a few things well. Malcolm Smith says '*When I arrived I could see things needed to be changed. It was pretty clear things were not being done terribly well. We had various groups building instruments for individual scientists on the staff. I went to the various places and asked people what was being done and it soon became clear that people did not know what people in the next offices were doing. They were in areas in which they could have helped each other and I realised that this was going to have to change*'. Change they did and as a result ROE built a reputation which it retained well after Malcolm Smith moved to Hawaii in 1985 and Malcolm Longair left the observatory for a chair at Cambridge in 1992. Matt Mountain, who would later build one of those cutting edge instruments, agrees that a key turning point came when Donald Pettie was appointed Chief Engineer and was then backed by Longair to convert the ROE from a traditional observatory into a proper, project-based organisation designed to be world class. Matt feels that an incredible innovative culture was created at ROE in the 1980s and 1990s and describes it, with 20–20 hindsight, as '*a truly amazing time*'.

## Dome Extension

It soon turned out that the minimum sized building specified by the original UKIRT design was in fact too minimal. This was no surprise to many, including Alex McLachlan who says that '*As far as the building was concerned the policy was to do what we had to do, but to keep in mind that it had to be possible to extend it along the ridge*'. Although the space below the observing floor was adequate, the small size of the original control room meant that the two PDP-11 computers and associated racks of instrumentation and interfaces occupied the majority of the space and contributed a high level of noise and heating to the general environment. If anyone was working on the computers, pulling out the racks to test or repair anything, it was very difficult to move around. The only advantage of the close proximity was that anyone who was working on the software or the computers could make changes while sitting at the telescope console. However, the heat in the room could become distinctly uncomfortable causing people to take extreme action. One UKIRT staff member turned up at the summit one evening to help and walked in to find a visiting astronomer sitting at the control console in his underpants, which he describes as '*Not a pretty sight*'.

The need for more space came to a head around late 1982 after which tenders were made, and quotes received, for a dome extension. Work was started during the summer season of 1983 and was largely compete by the autumn. The extension, which started from the little stub of the existing building just upslope of the roller door, allowed the removal of the computers from the control room, making it a quieter and cooler place for the observing team to work. The new extension was specifically designed to accommodate computer equipment and had its own air-conditioning and a halon emergency fire control system. The halon system was designed to dump its $5000 charge of halon gas should the smoke detectors pick up any sign of a fire. One day Sidney Arakaki and Charlie Richardson were in the control room when a recently delivered flask of liquid nitrogen that had been temporarily left inside the computer room blew its safety valve as it adjusted to the temperature in the room. This caused a cloud of vapour that triggered the halon alarm system to go active. With a 30 second delay before the halon dumped, and fire alarm horns blaring away, Sidney and Charlie had to rush from the telescope main floor, assess whether there was a real fire in the computer room and having concluded that there was no danger, get to the abort switch for the halon dump. The switch was fortunately very prominent. Charlie says he had '*Never seen Sydney move so fast, he beat me to the abort switch by two yards and beat the 30 seconds delay by seven seconds*'.

**Fig. 7.1** The foundations of the dome extension. L-R:- Alan Pickup, a visiting Astronomer, Alf Neild, Richard Wade, Sidney Arakaki (Photo Alf Neild)

About the same time as the dome extension was being built a new windblind was designed, although its delivery and installation had to be postponed due to financial constraints resulting from fluctuations of the $/£ exchange rate. The windblind was, however, given high priority by the UKIRT Users Committee and funds were finally allocated in late 1984. Actual installation would take a further year but, once it was fitted, it allowed the telescope to operate in winds of up 50 mph (80 km/h). Another advantage was that since two of the panels were made of Gore-Tex, which is transparent to long wavelengths, daytime observing in the sub-millimetre regions was possible without warming up the telescope structure to the detriment of the following night's observing. So, from 1984, routine daytime observing was offered at UKIRT with up to 30 % of the potential day observing time being opened up to UK astronomers, although in practice only a fraction of this time was ever used. The observations were limited to millimetre or sub-millimetre projects and the observing teams had to be self-sufficient, UKIRT was too short staffed to fully support both daytime and night time observing so in principle the daytime observers would be "on their own" in more ways than one. However in practice the night time telescope operators would sometime voluntarily stay on into the morning period and one remembers having to "force" the observers off the telescope about 11 am when the sun was getting too high, and him too exhausted, to safely continue. According to the January 1986 ROE bulletin this capability was soon put to good use, with the telescope being operated for up to 20 hours a day at times. It was not, however, a very practical arrangement. It was found that while morning observing could be done successfully, any afternoon observing had undesirable knock-on effects. Opening the dome in the late afternoon allowed in too much sunlight and the resulting heating of the structure led to poor seeing for the first few hours of the night. Soon sub-millimetre daytime observing was restricted to morning shifts and was heavily used by Ian Robson and his group from Preston.

**Fig. 7.2** The UKIRT control room soon after the dome extension was built is shown in this November 1984 photograph. PhD student Cindy Brown sits at the telescope control panel, the scientific instruments were operated and monitored from the rack at the left side of the picture (Photo Ian Robson)

## Flexible Scheduling

In June 1982 Tim Hawarden wrote a paper for the UKIRT Users Committee on multiple scheduling of the telescope. He suggested that two projects, one suited to normal weather and the other requiring good thermal or sub-millimetre conditions could be scheduled simultaneously and switched over depending on the conditions. Tim set out the problems of having two teams in Hawaii at the same time, the need to have one doing next to nothing while on standby and the tensions that would be created by one team having to give up the telescope because the weather was now too good for them. He also stressed the need for objective criteria and an impartial decision making process. To minimise some of these difficulties he proposed that the "takeover team", who would presumably be the sub-millimetre astronomers waiting for very dry weather, would remain in the UK and observe remotely if the weather improved. He even went so far as to speculate on triple scheduling with two groups on standby, one for the very best weather and the other one willing to use poorer conditions unsuited to the programme which was actually scheduled on the telescope on that night. This idea, while capable of maximising scientific productivity, found little support from the user community and was not adopted. About 10 years later, a version of this sort of scheme would re-appear as flexible scheduling, first on JCMT and later on UKIRT itself.

## Remote Observing

Tim's paper took into account a feasibility study of remote observing by Peredur Williams and Peter Thanisch in June 1981. In what was described as the beginning of a programme to make UKIRT observations by remote control, the telescope was first operated from the Hilo office on the 24th of February 1982. Communications for this first attempt were via two leased lines of 9.6 kbps capacity routed via the local telephone company's microwave system. One line was devoted to the slow scan TV system, which could download an image every 45 seconds to check that the telescope was pointed on target. The other line connected multiplexers at the summit and sea level, which allowed up to eight devices to share the bandwidth. This line was used to link the sea level operators with the summit telescope control computer and the instrument computer. Conventional telephone contact was used to ensure that any staff in the dome were not endangered by an unexpected movement of the telescope. As well as remote observations, the data link could be used for software development and data analysis.

Seen as a way of saving on travel costs and allowing scheduling flexibility, it was an experiment that would develop over a number of years. The plans were to gradually improve this capability, first with a microwave link to the summit and later, via packet switching networks, to the UK. This offered the possibility of remote operation of the telescope, or at least of its instruments, directly from the UK. The first such connection was made on the 1st of September 1982. This was followed a few days later by the first intercontinental remote observations with UKIRT when Malcolm Stewart, who had devised and implemented the system, observed the star HR8824 using a remote connection from Edinburgh. This achievement was reported in an SERC press release published on the 10th of September 1982 and was also used for publicity by the telephone company, GTE Hawaiian Telecom, who had provided the bandwidth necessary for the tests free of charge. Alan Pickup was at Hale Pohaku watching the Superbowl game on the TV when he saw an advert about the remote control of the telescope from the ROE, right in the middle of what is one of the most expensive adverting slots on American television.

By 1983 connections from ROE to UKIRT had been improved and plans were afoot to have a dedicated remote observing facility at the ROE that summer. The intention was not yet for full remote observing, but rather to provide the option for one or more team members to participate indirectly from Edinburgh while one of their colleagues made the arduous journey to Hawaii and back. Experiments with remote observing continued for a number of years but there were always some doubts about the justification for it. Was it a desire for improved efficiency, flexible scheduling or just the "relentless pursuit of technology"? Indeed was this a capability that the users actually wanted or was it just being pushed along by pressure from above? This first generation of remote observing experimentation continued until 1987 by which time the international packet switching service being used for data transmission had become congested to the point that speeds were too slow (it was often slower than a telephone modem) to be practical.

However, interest in remote observing did not fade away. SERC had set up a Steering Committee on the Remote Use of Overseas Telescopes which rejoiced in the acronym SCRUOT (although it was uncharitably known by some as the Steering Committee for the Remote Use of Telescopes Up Mountains, for which readers can work out the acronym themselves). SCRUOT commissioned ROE to produce a report on the financial and operational implications of operating UKIRT with a mixture of conventional visitor observing, observations made by local staff on behalf of absent scientists and remote observing. Peredur Williams took on this task and developed a model of mixed service, remote and classical observing in which several projects requiring different conditions would be scheduled together and swapped over depending on the weather. His conclusions were that despite the savings in travel costs being mostly swallowed up by extra staff costs in Hawaii, some monetary savings might be made and that unquantifiable benefits such as improved science quality and less time wasted in travel would be expected. He also pointed out the potential for lessons on the operational models for the soon to arrive 8 m telescopes to be learned relatively cheaply. Interestingly, notes from Tom Geballe and Tim Hawarden sent while this paper was being prepared are quite negative about the options being proposed.

**Fig. 7.3** Text based communication between the summit and remote observers in Preston who were involved in a blazer monitoring programme. The topic of the moment appears to have been the choice of biscuits (Photo Ian Robson)

A key to these models was better connectivity between UK remote observer and the telescope. The solution was to hire a dedicated data-line between Hawaii and the UK. This was requested as an "additional bid" in the ROE 1989 financial "forward look" and was ordered from GTE Hawaiian Telecom in the spring of 1990. After some delays resulting from high error rates occurring within the USA, the link finally came into service in March 1991 for UKIRT and in June 1991 for JCMT. The 56 kbps line went from Hilo to Honolulu, thence to California, New Jersey and, via the TAT-8 cable to the UK and then to Edinburgh, making it probably the longest private data line in the world at the time. No satellite links were used and reliability was fairly high. A dedicated remote observing room was set up at the ROE for the use of remote observers, the objective being to provide a reasonably well equipped facility, which allowed the remote observer to avoid the distractions of trying to operate from their normal office. Of course, since the ROE was part of the Starlink network, onward connections to individual Starlink *or* JANET (Joint Academic Network) nodes was possible if desired and remote observing sessions were held from a number of UK sites including Kent and Preston, but they never replaced visits by astronomers to the summit.

The dedicated line was a huge improvement over the packet switched network, especially for the UKT type instruments with their low data rates, but as the array instruments IRCAM and CGS4 came on stream, and particularly with the advent of 256 square arrays, it was always a struggle to keep pace with the ever rising data rates. Worse still, the dedicated line was expensive, costing $120,000 per year. Voice communications with the summit were originally via regular telephone line which, because of the expense, could not be kept open all night. Attempts were made to split off some of the bandwidth for an "always open" voice channel which was eventually achieved as far as Hilo but never worked satisfactorily when extended to the summit. This was perhaps the most frustrating thing about these early experiments, when things got hectic and the remote observer wanted to know what was happening, the observing team at the other end was usually too busy to answer the phone. Worse, since this was well before the days of cheap web-cams, the remote observer often had no idea if the summit team were in the control room, or even in the building. Attempts to improve voice communications by using compressed voice links multiplexed with the data stream were not entirely successful since they introduced other network reliability problems.

**Fig. 7.4** GTE Hawaiian Telecom was proud of their part of the remote observing project. Readers familiar with the ROE site will notice that the Edinburgh picture has been left-right flipped to make a more pleasing image of the curvature of the two summits (Via Peredur Williams, with the consent of GTE-Hawaiian Telecom)

In the end, despite the arrival of the internet which made international data transmission so much easier and cheaper, the UKIRT remote observing experiment turned out to be just that. The costs of the dedicated line proved too much, or the savings being made on travel and time too nebulous, for the project to continue. In particular the travel costs, which in the 1970s had been rising and looked likely to continue to do so in the light of oil shortages, stopped increasing and today you can fly to Hawaii for almost the same price as was possible two decades ago. Adding to this was a lack of enthusiasm for remote observing in Hawaii, where the management worried about the local staff being isolated from their user community if the telescope went down this road for routine operations and the need to provide extra staff at the summit for safety reasons when no visiting observer was to be present. The lease of the dedicated line was not renewed when it fell due in March 1993 and, although network connections still existed, they were not as fast or as dependable as the dedicated line and the remote observing project was wound down. Nevertheless, these early tests were extremely valuable and showed that many of the problems were social in nature, for example the inability to communicate in real-time, rather than in the technical nature of the observing itself or the data flow back to the UK.

In due course weather dependent flexible scheduling would come to the JCMT when Ian Robson, a keen exponent of the principle of remote observing, became its director. He eventually introduced weather-based, queue observing with remote data viewing, which eventually spread to UKIRT and to a new generation of 8 m telescopes. However, the contribution of these early UKIRT experiments is rather difficult to quantify. They clearly set people thinking about different ways of doing things, but remote observing was probably a few years ahead of its time and when the UKIRT experiment ended, few mourned its passing.

## Service Observing

Sometime about 1982 Andy Longmore proposed a programme of short observations carried out by local staff for rapid follow up of discoveries from the soon to be launched IRAS satellite. This programme was originally called, logically enough, IRASserv. According to Andy the idea appealed to Tim Hawarden, then based in Edinburgh, who suggested expanding the scheme to other, non-IRAS, programmes requiring short (two hours or less) observations that would not merit a dedicated observing run of several nights. Thus the UKIRT Service Observing programme, UKIRTSERV was born. Six nights of time were awarded to the service programme for the second semester of 1983 and a further eight nights were allocated in the first half of 1984. At first not everyone was convinced this was a good idea. For example, Bob Joseph asked for photometry of an object at the position of one of the sources reported by the IRAS satellite to confirm if the IRAS source was an interacting galaxy whose position was nearby. The report from the UKIRT service observing came back that the colours were those of a star, but a few months later Joseph's team was observing at UKIRT and they confirmed that the IRAS source was indeed the galaxy NGC 6240 after all, a result which became one of the celebrated early discoveries from the satellite.

Despite these sorts of glitches, in 1984 the service programme was given long term status so that blocks of time were set aside every year automatically. Applications were solicited by e-mail using a simple text-based form every 8 weeks or so and these were quickly refereed by one or two UKIRT staff and a single external astronomer. Approved observations were then placed on a list and done on one of the scheduled service observing nights or whenever a few hours of time presented themselves, for example during engineering nights. Tom Geballe was a key part of the service programme at the Hawaii end, being involved in it as both referee and observer throughout his time at UKIRT and for more than a decade after he left UKIRT to move to the Gemini Observatory.

Andy Longmore returned to Edinburgh in 1985 and brought the running of the service programme with him. He continued to operate the programme until the transfer of Richard Wade from ROE to Hawaii to become Director of the JCMT in August 1989 caused some changes in the UKIRT team in Edinburgh. Longmore became Head of the UKIRT Unit and John Davies, who had left British Aerospace a year or so after hearing Ian Robson's talk and had arrived at ROE in 1987 via Leicester and Birmingham Universities, was assigned management of the service programme. John was happy with the change, it provided an opportunity to make occasional observing trips to the telescope and develop his observing experience which had lapsed after several years of laboratory based work on space projects.

**Fig. 7.5** Andy Longmore (Photo ROE)

After John moved to Hawaii in 1993 the programme passed to Suzie Ramsay who oversaw the 1000th application in September 1996 before responsibility for service observing was returned to Hawaii as the UKIRT unit at ROE was phased out. In Hawaii John took up the reins again for a while before handing over to Paul Hirst, who replaced the traditional text form with a more modern web-based interface (a prototype of this web-interface had been presented at a WWW conference in May 1995 by Henry Stilmack, a member of the JAC computer support staff). The service programme continued to be successful and ran almost until UKIRT operations changed to virtually full-time survey mode in around 2009.

## IRAS Follow-up

Service observing was not the only means employed at UKIRT to try and exploit discoveries made from the IRAS satellite. Follow-up observations of IRAS sources had been approved by the UKIRT Time Allocation Group and it was planned that Ian Robson's sub-millimetre group would operate in the morning daytime slot, attempting to do rapid follow-up on IRAS sources. Unfortunately, this was not a success and the run was abandoned part-way through because it turned out that many of the IRAS sources had large positional uncertainties, so could not be found, and the weather was bad. It was also clear that on balance the idea not cost-effective and the observers found that being trapped at Hale Pohaku for days on end with access to only three hours of telescope time was not a good use of their time. The Netherlands had also participated in providing an IRAS follow-up instrument, called logically enough IRASFU, although its lack of success (only one published paper can be easily identified) led to some other translations of the acronym. It was a scanning array device, that was novel in practice, but unfortunately there was no data analysis software and so trying to understand what was happening at the telescope in real-time was a huge challenge and the instrument did not have a long lifetime.

## The First Telescope Operators Arrive

Although a plan for up to five night assistants had been agreed in principle in 1974 it was not until early 1983 that UKIRT hired its first such staff, who would later and more accurately be known as Telescope Operators (TOs). The need for this role had become more critical as the existing UKIRT support astronomers found themselves increasingly overloaded, particularly at weekends when there was no engineering "day-crew". Andy Longmore gave an example of what it was like at the time. He would rise around noon and, after what was breakfast for him and lunch for most of the other people around, would drive up to the summit to fill cryostats and carry out any other maintenance or essential running repairs. After completing these tasks he would return to Hale Pohaku for dinner, then immediately drive back up again for a potentially 14 hour night supporting visiting astronomers. Driving back down at dawn he would get only a few hours sleep before having to start again. After a few days and nights of this he was exhausted. Providing dedicated night-time support staff broke this cycle and allowed a more sensible working pattern for everyone. According to Bob Joseph '*During these first years the support astronomer would teach the visiting observers how to run the telescope and after a night or so would turn the telescope over to them. This made for rather inefficient use of the telescope, and it was a huge boost to the scientific output of the telescope when telescope operators came on-board*'.

The first two telescope operators were Dolores Walther, who joined on the 23rd of February, and Jack Pacleb. Dolores says that she '*Used to sit in my classes and look at the mountain wishing that I could work up there. So when I graduated I was planning on going to the University of Hawaii at Manoa to graduate school to get a degree in astrophysics. As it turned out UKIRT advertised for a telescope operator position and because I wanted to work up on Mauna Kea I thought I would apply and I got the job. Hence I was the first telescope operator they hired* '. She was interviewed by David Beattie, Terry Lee and Tom Geballe and, some time after she was hired, Tom told her, "*I didn't want to hire you, I wanted Jack because I knew we had you and I was afraid we would lose Jack*". According to Terry Lee the interview result was a tie and so both were hired. Ironically Jack Paclab didn't stay for long, he left in late 1984 to become a policeman. There was a clearly an urgent need for a replacement and from 123 applicants Joel Aycock joined UKIRT in early 1985. Another operator, Stephanie Salazar was also hired but she was soon replaced by Rick Roper, who himself left in the summer of 1985 when he found himself unable to cope with the headaches caused by working at altitude. In August Thor Wold, whose c.v. included a spell as a disc jockey, was recruited to fill the gap. Joel Aycock remained at UKIRT for several years before he moved to a similar position at the Keck Telescope and was replaced by Tim Carroll. The resulting team of Dolores, Thor and Tim formed a highly professional team and became the backbone of the night time observing at UKIRT for the next decade.

**Fig. 7.6** Telescope operators Thor Wold, Joel Aycock and Dolores Walther at Hale Pohaku (Photo Dolores Walther)

## Isaacman's Bend

Despite the near death experience of the commissioning team near the summit, driving remained both a popular sport and perhaps one of the biggest risks faced by the astronomers. One UKIRT staff member, now a senior academic, admits that when he was supporting a visiting observer, also now a senior figure in British astronomy, the pair engaged in some driving competitions. They were not racing, but rather competing to see who could make the morning drive from the summit to Hale Pohaku the fastest. On one such time trial the driver almost lost control and looked to be coming off the road into a field of large boulders, any one of which could, as he put it, have '*Totalled the car and probably its occupants*'. After this near miss the pair agreed by mutual consent to stop the competition. It was thus ironic that a spectacular crash then occurred in relatively benign conditions; in daylight and on a dry road.

Rich Isaacman had "borrowed" a UKIRT vehicle to carry an astronomer friend and former university roommate, who was visiting Hawaii and was interested in seeing the summit. On the way down from the mountain, as the car rounded one of the hairpin bends, Isaacman started to straighten the wheel, but the car completely failed to complete the manoeuvre. The power steering drive belt had jumped off its pulley, which of course made the steering suddenly much, much stiffer on such a heavy vehicle. Caught by surprise, he did not apply enough force to turn the wheel and the car continued to turn and flew off the road to the left. It then rolled over, and landed upside down in the roadside ditch. He feels that '*There is no doubt that we were spared serious injury by the roll bar since the vehicle collapsed slightly under its own weight, shattering the windshield, and would very likely have crushed us without the reinforcement. As it happened, we were stunned (and hanging upside down in our seatbelts) but uninjured*'. His passenger John McCarthy clearly remembers '*hanging from my seat belt, upside down, looking out the windshield. Suddenly it crazed, crinkle-crinkle-crinkle, and I figured it was time to get out. I released the seat belt, fell to the roof, and crawled out the side window. Rich and I were both unhurt, a little shaken up, I guess. We quickly realized we were lucky that the truck had gone off the road on the uphill side. If we had gone over the downhill side, there would have been nothing to stop us from rolling, and rolling, and rolling..*'. The two astronomers hiked down to Hale Pohaku, where Ian Gatley reassured them and phoned Peter Forster, who came and fetched them.

**Fig. 7.7** Rich Isaacman (in the white sweatshirt) and John McCarthy examine the Ford bronco at Isaacman's bend (Photo Rich Isaacman)

There was an internal safety enquiry into the accident, Alan Pickup remembers '*going up after the accident and measuring the skid marks. Then driving the vehicle and braking hard to see what skid marks we left so that we could work out how fast the car was going when he left the road*'. Although Isaacman had taken the vehicle for his day trip without asking anyone (it was not uncommon for people to do this) no official action was taken apart from him getting a lecture from the safety officer, but inevitably the corner became known for years afterwards as Isaacman's Bend. This was, alas, not his only run-in with Ford Broncos. On a later occasion, while driving a visiting astronomer from Holland up the mountain at sunset for an observing run, he became blinded driving straight into the setting sun and ran off the edge of the road, bottoming out the vehicle and causing some damage there.

Kevin Krisciunas was also witness to a Bronco rollover when some visiting engineers or astronomers working at the CFHT were racing down the mountain to Hale Pohaku, allegedly trying to see if they could beat their own record time for the trip of 12 minutes, which would require an average speed of over 40 mph (65 km/h) on what was then an entirely dirt road. At the 11,000 ft level they hit a bump, lost control and went off the left side of the road. Kevin drove the summit ambulance down to the accident and waited for Army medics to arrive.

**Fig. 7.8** A CFHT vehicle a few days after an accident. One such rollover occurred in January 1985 according to the ROE bulletin of February 1985 (Photo ROE archive, origin unknown)

## Unwanted Tenants

Sleeping over in the UKIRT dome was prohibited for health reasons but from time to time the building did host temporary residents, usually mice who had somehow made the trip up from sea level and needed somewhere warm to live. Ian Robson '*Never knew how they got there but nevertheless mice and electrical cables do not mix and so a concentrated campaign was undertaken to eradicate them, much to the dismay of some of the staff who were caught in two minds about the plight of the mice and the plight of the telescope. The mice were either totally fearless or had become very human tolerant because one night I felt this movement in my cold weather jacket draped over the back of my chair and stuck my hand in the pocket to find a mouse had climbed up the sleeve and was starting to attack my packet of sandwiches. Needless to say, both I and the mouse were somewhat surprised and the mouse jumped out and fled across the control room before I could throw anything big and heavy at it. Dolores wasn't sure who to shout at, but she was sure relieved I missed the mouse*'.

## New Facilities at Hale Pohaku and the Summit

Since much of Mauna Kea was (and still is) a protected area, various development plans had been put in place in the mid-1970s and so permits were required before any work could be done to improve the astronomical infrastructure. This in turn led to the need for a master plan for the summit area which had to address the needs of all the interested parties including conservation groups, the army training base in the saddle area, hunters and the public. As the host institution for the Mauna Kea Observatories, the University of Hawaii was responsible for producing this plan. This was not a trivial matter as there was opposition to any development on Mauna Kea through a mixture of genuine concerns, conflicting views and lack of understanding of the aims of the astronomers. Among the latter was the incumbent Mayor of Hawaii county who allegedly compared the view of an observatory building from Hilo to a pimple on the mountain. However, according to Terry Lee there was also political support from within the County Council and its administration.

As part of the continuing education programme at Hawaii Community College in Hilo Professor Mary Matayoshi (the Mayor's wife) led the Mauna Kea Forum, a series of public meetings that discussed the various interests of tourism, conservation, native Hawaiians, recreation, hunting, astronomy etc. There was no real conflict with the hunters, there was no game in the summit area and the safety zones around the road and mid-level buildings were not to increase substantially, so hunting was largely unaffected. Initially the conservation issue centred on the Palila bird, one of the few native species not eradicated by the Hawaiians in their harvesting of feathers or by the rats imported by visiting ships. The bird was considered to be an endangered species and it was the subject of a continuing study. During the course of the work the study concluded that the influence of activities at Hale Pohaku had a positive impact on the bird population. Furthermore, the main threat to the Palila was the mouflon (wild sheep) which grazed on the young trees whose seeds formed the main part of the bird's diet. The upshot was a recommendation to eradicate the mouflon, which did not please the hunters in the long-term. Later, one arm of the conservationist movement expressed concern for the Weiku Bug, a wingless fly, as being endangered by construction in its summit cone habitat. However, at the time the main concern of the native Hawaiians was the protection of their sacred sites, the principal of which was the cinder cone Puu Poliahu, while access was a concern of those who feared that the road might be closed to the public. As part of the process of informing the public some talks on astronomy were arranged. Ian Robson gave one such talk which was apparently very good, but sadly not well attended. Bill Zealey was also responsible for some public relations work and spent some time '*Roaming around the island showing SRC and other astronomy films*'. One day he learned that it pays to view such films privately before showing them in public. During one event at which he stressed that astronomers were environmentally sensitive people he showed a documentary about the Multiple Mirror Telescope in Arizona. Three minutes into the film there was a scene in which the construction crew blew the top off their mountain site in a violent explosion!

Before its approval by the University Board of Regents in January 1982, the Mauna Kea Research Development Plan went through its various stages, public hearings, the submission of an Environmental Impact Statement and clearance by the Department of Land and Natural Resources. This last step was not trivial since one of its staff members was felt by some to be as obstructive as he could get away with. As part of its processes the Department of Land and Natural Resources held public hearings before making its recommendations and the first of these which Terry Lee attended was for the California Institute of Technology's proposed sub-millimetre observatory. The meeting was in Waimea and Terry drove over to the event with Helene Hale. He thinks that it may have been on this occasion that it was mentioned that land at the Hilo Bay Front had been designated for use as soccer fields. Someone said "You British guys play soccer don't you. Maybe you should help with coaching the kids". This simple activity was another huge step towards integrating the astronomers in Hilo into the local community.

## Hale Pohaku

The long term provision of facilities at Hale Pohaku was a part of the Mauna Kea infrastructure project, which also concerned the road and the electric power line to summit. For the mid-level facilities a set of general specifications existed as early as February 1978, but John Jefferies describes the history of actually getting the buildings approved as '*tortuous*'. He says that while he '*put a great deal of thought and care into their design he suffered a lot of abuse from local environmentalists, from some of the Hawaii bureaucracy, and a few of his IfA colleagues*'. He wanted buildings that would be compatible with the world-class astronomy programme he was sure would emerge on Mauna Kea, but which would also be a haven for the people who were working in the extreme conditions on the summit. He wanted the facility to be '*a source of pride and pleasure not only to the people of Hawaii, but to all who came to use the mountain*' and recalls that all involved had to fight to achieve this, often against people who complained that the buildings were unjustifiably lavish. The design was done in close collaboration with the other telescope teams and a working group was formed at which representatives from the telescopes met with the architects. Andy Longmore represented UKIRT at these meetings.

The agreement signed by SRC had anticipated paying its share of accommodation costs, the daily amount was well within what was by now SERC's subsistence rates and so could be fitted within current budgets. So agreement was easily reached on running costs, although SERC wanted a provision to buy out the capital element of the new mid-level buildings should they find themselves with cash to spare. The various observatories were asked to indicate the number of rooms they were likely to need and Terry Lee signed up for what was about the maximum UKIRT might need in the case of 24 hour observing at rather than an average number. This over-estimate would work out well when JCMT operations started in 1987.

Peredur Williams remembers an interesting aspect to the construction work concerning the provision of sanitation which, on the Big Island of Hawaii, often relies on cess pools rather than conventional sewers, especially in remote areas. The usual practice was to blast holes in the ground with explosives until a suitably large natural lava tube was found in the rocks and the sewage could be diverted into it. To allow the night observers at least some rest, blasting was delayed until after 1 pm at which point a whistle would sound followed by an explosion about 15 seconds later. Inevitably this blast would wake everyone trying to sleep in the flimsy original buildings.

The new Hale Pohaku buildings were dedicated on the 11th of August 1983 at a ceremony attended by the Mayor and other local dignitaries. The new complex featured a main building with an improved cafeteria, offices and laboratory space for the various telescope teams and their visiting observers. A large room containing shelves of books and star atlases was referred to as a Data Preparation Room since the planning consent did not allow the provision of a library except at the various sea level headquarters. Several accommodation blocks were provided, one for resident Hale Pohaku staff and two (later three) for visiting astronomers. From a visitor's point of view a key improvement was the provision of better rooms each with a private toilet and shower and the ability to separate day- and night-shift workers, improving the likelihood of a decent sleep for all. As well as far greater comfort, the new facilities had much more space for study and recreation. The pool table in the old camp was always busy and by popular UK demand the new facilities were to have a snooker table in addition to a large size pool table. The evening before the dedication of the new facilities there was a dinner for senior members of the participating organisations. John Jefferies and Terry Lee played the inaugural game of snooker on the new ¾ size table. They were equally inept, the game never finished!

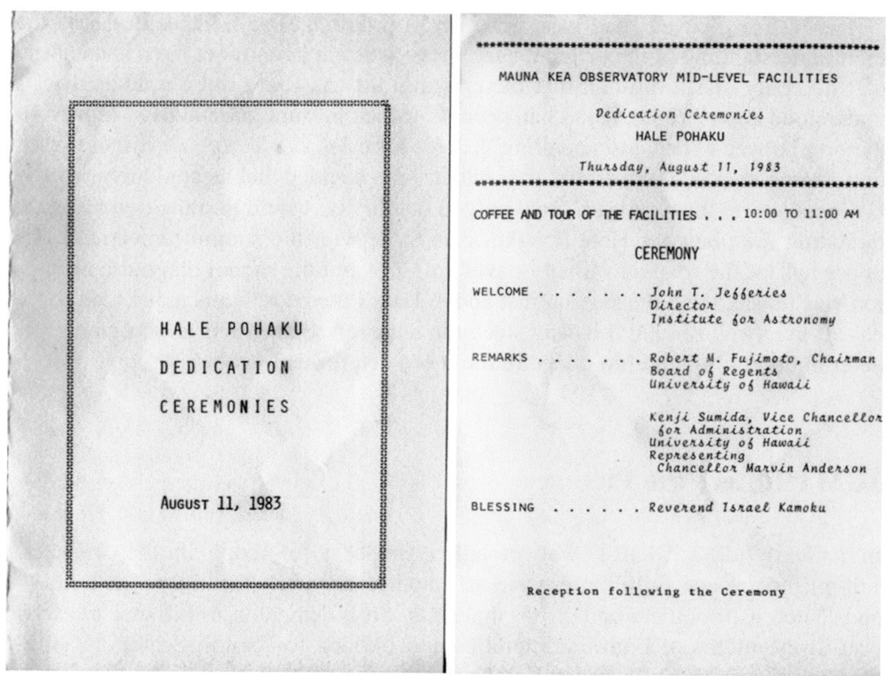

HALE POHAKU

DEDICATION

CEREMONIES

AUGUST 11, 1983

MAUNA KEA OBSERVATORY MID-LEVEL FACILITIES

*Dedication Ceremonies*

HALE POHAKU

*Thursday, August 11, 1983*

COFFEE AND TOUR OF THE FACILITIES . . . 10:00 TO 11:00 AM

CEREMONY

WELCOME . . . . . . . *John T. Jefferies*
*Director*
*Institute for Astronomy*

REMARKS . . . . . . . *Robert M. Fujimoto, Chairman*
*Board of Regents*
*University of Hawaii*

*Kenji Sumida, Vice Chancellor*
*for Administration*
*University of Hawaii*
*Representing*
*Chancellor Marvin Anderson*

BLESSING . . . . . . *Reverend Israel Kamoku*

Reception following the Ceremony

**Fig. 7.9** Bill Parker preserved his copy of the programme for the dedication ceremomy (Photo Bill Parker)

## Power to the People

The power line project was intended to provide mains power all the way to the summit. This was to be done by connecting Hale Pohaku to the power line running through the saddle between Mauna Kea and nearby Mauna Loa via an overhead line, then taking the cables underground to the summit. This supply would replace the diesel generator plants at Hale Pohaku and near the summit. However, improving the power supply and the road to the summit would require a considerable sum of money and it would not be forthcoming from the University of Hawaii. For new projects proposing to come to Mauna Kea, buying into or expanding the infrastructure would become a normal part of the terms of any agreement, but in the case of existing projects (for example UKIRT, CFHT and IRTF) terms compatible with the existing agreements for each of the telescopes and practical constraints needed to be worked out. For UKIRT the task fell on Terry Lee. Analysing the numbers he found that the cost of electricity via the proposed power line worked out a bit less than what UKIRT was paying for the diesel generated variety so it was reasonable to carry on paying at a properly audited rate.

Paying for a paved road was, however, a problem. The UK position was that their understanding of the original agreement was that a road was the responsibility of University of Hawaii and that the existing dirt track was not a road as the UK understood it (in 1979 the road had been described in some official SRC minutes as varying between "bad and appalling"). John Jefferies and Terry Lee discussed this one afternoon over a few drinks after which John decided that he could live with the UK position on the road and Terry judged that SERC would find the overall terms, including the plans for Hale Pohaku, consistent with the summit agreement. The upper half of the road was finally paved in 1989, but the money ran out before the job was finished and the section just above Hale Pohaku remains a dirt road to this day. It may well remain this way since the unpaved section acts as something of a deterrent and limits tourist traffic to the more determined visitors.

## JCMT Enters the Picture

In its early days UKIRT was probably the best telescope in the world for sub-millimetre and millimetre wave astronomy, research areas in which both UK and Dutch astronomers had strong interests. So a dedicated millimetre telescope was in the minds of both communities and options for building such a facility, perhaps as part of the Northern Hemisphere Observatory project in La Palma were beginning to evolve. Eventually the superiority of Mauna Kea for sub-millimetre observations decided the issue and the location of the new telescope was switched to Hawaii. It would of course be operated from the same base as UKIRT, eventually leading to the expansion and renaming of the UKIRT sea level base as the Joint Astronomy Centre.

By mid-1983, as the master plan was being signed, work was already under way on the millimetre-wave telescope building and its foundations were almost complete. Progress was sufficiently rapid that by the end of 1984 the sub-millimetre news items previously reported in the UKIRT Newsletter would be moved to a dedicated publication called Protostar. Staff changes were also afoot. Phil Williams began a tour of duty in Hawaii in January 1984 and in April it was announced that by the end of the year, the conclusion of his tour of duty, Terry Lee would return to ROE as Head of the Technology Unit. It seems that to help bring about the changes he was trying to make, Malcolm Longair wanted Terry's recent operational experience and wide technological expertise back in Edinburgh. In what was a straight swap of roles, he was to be replaced by Malcolm Smith as Astronomer-In-Charge. Malcolm would be responsible for both UKIRT operations and for the build-up towards the operations phase of the millimetre-wave telescope. David Beattie would remain as Malcolm's deputy. On the technical side Mark Horita replaced UKIRT technician Mike Santos who left to rejoin the US Navy.

1984 also marked the decommissioning of UKIRT's f/9 focus and the f/35 focus became the default top-end for the next 20 years. However, the f/35 secondary did have one undesirable feature, it was originally[1] powered by a hydraulic system so hydraulic lines, or more precisely *leaky* hydraulic lines, snaked around the secondary mirror mounting. As a consequence, the UKIRT primary mirror received an occasional very fine spray of hydraulic oil. Fortunately the oil's infrared emissivity was low and therefore it did not affect observations too badly, so the astronomers just put up with it; or at least, most of them did. One visiting astronomer, whose name is providentially forgotten, was quite upset by this and, when no one was looking, took it upon himself to improve his signal-to-noise ratio by polishing the primary mirror with a cloth. Needless to say, all he succeeded in doing was smearing the oil around, and, more dramatically, removing the aluminium coating within the sweep of his arm's length. He was found out and forever banned from the observatory, but for a couple of years afterwards it was possible to defocus a star image on the visual monitor and see a little semi-circular bite taken out of the edge from where the mirror was no longer reflective.

**Fig. 7.10** David Beattie and Malcolm Smith in 1985 (Photo Susan Beattie)

## New Instruments

In June 1984 David Beattie and Ken Maesato visited the ROE to take part in the final testing and acceptance of a new InSb photometer called UKT9. Basically an improved version of the earlier photometers UKT5 and UKT6, it had been built with the lessons of these earlier models in mind and offered lower noise and improved stability, and so greater sensitivity, than its predecessors. UKT9 arrived at UKIRT in July and would be a workhorse instrument for a number of years, surviving until 1995, well after the arrival of array instruments like IRCAM, which would eventually replace it. UKT9 was often used with a large aperture enabling it to observe low surface brightness sources; such as extended emission from

[1] Replacement of the secondary mirror hydraulic system was one of the Dutch contributions mentioned in Chap. 6.

molecular hydrogen gas, and it could also be used with a Fabry Perot Etalon mounted directly in front of the dewar. About the same time the eight channel thermal infrared bolometer UKT16, known colloquially as the “8 banger” became operational. This instrument, designed and built by Ian Gatley and Alf Neild, employed eight detectors in a four by two array and could be operated with filters which approximated the M, N, Q and 30 μm filter bands.

These instruments could be mounted on a new Instrument Support System (ISU2) that was commissioned in August 1984. ISU2 was able to mount a number of instruments simultaneously and it was stiffer than its predecessor. This solved one of the annoying features of the “Gold Dustbin” which was its failure to maintain instrument alignment during the night. This was because the original unit had been designed with quick-disconnect brackets to attach the various instruments to the telescope. It turned out that instruments sagged during the night, so that it was impossible to do photometric standards around the sky and get back to the same magnitude on the same star later, because by then the instrument alignment had changed.

ISU2 offered greater observing efficiency, especially since its movable dichroic mirror could be computer controlled, making switches between instruments during the night quick and repeatable. Ian Sheffield, now back in Edinburgh, worked on the ISU2 software. In fact, he worked on it twice since the original telescope system was controlled by a PDP-11 computer with quite a tight operating system which meant it could do real time operations without having to worry about variable response times. So at first ISU2 was controlled directly from the PDP-11. Later, when the computer system was changed, ISU2 needed subsidiary electronics to control the mirror. So ISU2 was redesigned to have a micro-processor of its own to detect when the dichroic mirror was approaching its desired position and when it had to stop moving.

By mid-1984 the instrument complement looked like this:

| | | |
|---|---|---|
| UKT5 | J, H, K, L, L’, M filter photometry | Temporarily decommissioned and replaced by UKT9 |
| UKT6 | J, H, K, L, L’, M filter photometry + CVF | |
| UKT7 | L’ to Q filter photometry | |
| UKT8 | N, Q filter photometry + 10 μm CVF | |
| UKT9 | J, H, K, L, L’, M filter photometry + CVF | |
| UKT10 (The 2 Banger) | J, H, K filter photometry | |
| UKT16 (The 8 Banger) | L, N, Q filter “mapper” | 4 × 2 detectors |
| IRASFU | N, Q band filter “mapper” | 16 detectors |
| CGS2 | Near IR spectroscopy | 7 detectors |

In mid-1985 a sub-millimetre common-user instrument called UKT12 was being commissioned at UKIRT, but it looked like being short-lived since by September 1985 a new sub-millimetre photometer (UKT14), was nearing completion. When it was ready Bill Duncan and his family moved with it to Hawaii and remained there for a three year tour. UKT14 was commissioned on UKIRT in January 1986, but with the millimetre-wave telescope (JCMT) soon to become available it was planned that this instrument would be shared between UKIRT and the JCMT. In fact UKT14 moved to the JCMT permanently in February 1988.

## Surprise Visitors

In later years, especially after half of the road above Hale Pohaku was paved, the summit of Mauna Kea became a popular tourist destination. By the late 1990s van loads of tourists were being brought up the summit almost every night and the area round UKIRT on the summit ridge was a favourite stopping point for these quite respectable visitors. Earlier visitors were both fewer, and rather less respectable. Alf Neild met one such group

*'Rory and I had just brought some gear in to the dome, through the roller door; and had left it about 30cm up from the floor, to let a bit of cold air in to help reduce the dome thermals. Suddenly, a male body rolled under the roller door; dressed in a thin "T" shirt, shorts and flip flops. (Outside was snow covered, and was at -5°C.), He was followed by another 4 males; similarly dressed. After "Humouring" them, and letting them use the bathroom; we managed to get them out of the dome, and lock up. On the way down, we met four of them again; in the valley between the top cone; and the perfectly shaped cinder cone. I asked where their friend was, and they pointed up towards the cinder cone rim. Their friend was at the top, and sat on a surfboard: ready to launch himself down the 45 degree slope. I said "Tell your mate to walk down. He will kill himself, if he uses the surfboard". To which his mate said, in a slurred voice, "My bro got the balls and the guts to surf down". To which I replied, "If he does, you will probably see them both; on that lava outcrop". We then drove off'.* Alan Pickup also encountered an intrepid snowboarder who had gone one step further. He says *'On one occasion, I think it was New Year's Day, we got a knock on the door and it was the girlfriend of a guy who had tried to snowboard down the summit on a boogie board. He had careered down and when the snow stopped he didn't. He finished up bouncing down along the rocks and cinders at the bottom, painfully. We got involved in putting him on stretcher and getting him evacuated. We had to get the stretcher and the oxygen out. But of course it's very hard to get someone up a steep slope on a stretcher when for every step up you just slide down again. In the end we got assistance from the Army who sent up a helicopter to recover him'.*

Another legendary, albeit not so bloody, event was described by Dolores Walther as follows '*One night we were doing CVF work and while watching the strip chart* [*on which the signal was displayed at that time*] *we noticed that there was something very strange going on. I believe Andy Longmore was up at the summit at the time. I looked around outside for clouds but nothing, we couldn't explain it*'. As Andy and Dolores continued to puzzle over their data the explanation came from the telescope operator from the NASA IRTF who telephoned to say he could see said a "gentleman" (not his actual word according to Andy) stoking a fire under the dome slit. Apparently he was trying to keep warm, not realizing that the smoke was causing problems for the astronomers inside.

More remarkable than these macho males however have been the near legendary stories of uninvited female visitors. There are at least three stories of such visits and they are so contradictory that one can only assume that all three refer to different incidents.

Late one afternoon Ian Robson opened the door to find a woman in a one piece garment that might be politely described as diaphanous. He recalls thinking '*She hasn't got any shoes on*'. This was perhaps not the effect on which the lady was planning, but Ian recalls that there was snow on the ground and the temperature was close to freezing. She was not invited in and, according to Ian '*We never did find out what happened to her*'.

However, there does not seem to be much doubt about this incident, recalled by Bill Parker and Kent Tsutsui, which was apparently recorded in the engineering day book.

'*I remember I was in the control room when I heard a knock at the door. I opened the door and there was this naked girl, not very old, with a clothed younger male in attendance looking a bit embarrassed. She asked to come in and I advised her to clothe herself against the sun, but of course she was an untutored person and didn't think my advice was worth much. I think she put some minimal thing on. I led them in and attempted to explain what the telescope did but I could see from the blank stare that they didn't understand any of what I was talking about. I then led them into the control room and Kent's eyes nearly popped out of his head, he was so flabbergasted at seeing a naked (or nearly so) girl in such a situation. By the way, I remember she stated quite brazenly that she was a "hooker" – her word – and what her charges were. After they had seen the control room I think she used the bathroom and then I showed them out. I don't remember the young guy speaking at all while they were there*'. Kent adds that shortly after Bill had gone to open the door he '*Saw some movement through the control room window and saw a topless girl and a guy with Bill. He brought them into the control room where I stammered a "What's going on?" She answered, "Aren't you enjoying this?" I think I answered "Hell, yes but we're working here."*'

Bill then took them out of the control room and downstairs and Kent sat down, trying to collect his thoughts. Suddenly there was more knocking at the side door. This time it was an older gentleman in cold weather gear who began to ask about applying for observing time on the telescope. Before Kent could answer, he heard Bill and his companions coming up behind him at the top of the North stairwell. The elderly visitor took a quick glance over to them just as the girl was giving Bill a thank you kiss. Without a break in his request he then continued on about his qualifications as an amateur astronomer in a completely serious manner. Kent says

*'My mind began to boggle at what was going on around me and just standing there politely listening to the gentleman was all I could do from bursting out laughing'*. Eventually, they all departed the summit. Bill admitted to Kent that the girl was naked when the couple arrived but that he got her to put her shorts on and Kent "*communicated his disappointment*" to that. The two laughed a bit more and then decided to call it an early day and go home.

**Fig. 7.11** Kent Tsutsui waiting for the road to the summit to be cleared of snow (Photo Alf Neild)

Chris Impey and adds another, which seems different again '*There were also hookers and a pimp up one night I was observing, it was late fall and bloody cold and they were in a jeep and tank tops and flip-flops. We of course wouldn't let them in the dome, so one of them started lobbing stones in the slit, but they got bored and left*'.

## Red Alert

After Peter Forster had produced his list of altitude sickness "Red Alerts" they were printed on little cards and given to staff as a reminder of what to look out for at the summit. From time to time they did just what they were supposed to do. One cloudy night Rich Isaacman and Ian Gatley were at the telescope, waiting out the weather and hoping against hope that the conditions would improve. In the middle of the night they received a phone call from an astronomer having a similarly unproductive night at one of the other telescopes who asked if he could come over and borrow an amplifier manual. The visitor arrived a few minutes later, complaining about feeling awful and of having a terrible headache. Altitude related headaches are common on Mauna Kea, so at first the UKIRT astronomers just made appropriate sympathetic noises. Then the visitor remarked that he also felt a strange internal pressure behind his eyes, as if they were going to pop out of his head. This made Gatley and Isaacman sit up and take notice since, thanks to the red alert cards they carried they immediately recognized this as a symptom of cerebral oedema (a swelling of the brain caused by altitude). They asked "*Anything else*?" and got the reply "*Yeah, my vision's all strange, dark at the edges like I'm looking through a tunnel*." At this point the two of them jumped to their feet and said, "*OK*! *Time to go*! *Right now*!" They bundled their visitor into a Bronco and sped downhill. The ailing astronomer wanted to stop to pick up his things but Ian Gatley would have none of it, he did not want to waste any time whatsoever. Isaacman got out of the car at Hale Pohaku and Gatley continued down the mountain directly to the Hilo Hospital. They kept the sufferer overnight and released him, fully recovered, the next day.

Charlie Richardson had a similar opportunity to use the first aid knowledge imparted as part of the staff's medical training. An astronomer who had never been up to the summit before complained of shortness of breath and of not feeling well. Richardson, Pat Nelson and Ken Maesato examined him more closely and Charlie recalls that they '*could literally hear popping sounds as he breathed, a sound like Rice Crispies*'. Knowing from their first responder course exactly what this indicated (pulmonary oedema or fluid in the lungs) they had no doubt about the need to get him down the mountain quickly. They found out later that their visitor had been exhibiting signs of pulmonary oedema and that they had indeed taken the right action.

Forster had also warned that children are most susceptible to altitude sickness and was vocal in discouraging people from taking them to the summit. This led to some friction with the local school district's summer science enrichment programme, one of whose features was a field trip to the summit. Isaacman was one of the observatory volunteers who would give a talk, then accompany the group up the mountain for a tour of UKIRT, but every year the same thing would happen: some children would get sick and pass out. After two years of this, when the school district contacted him about helping out for another summer's programme, he told them that he was happy to give the lecture but that he would not take the children up the mountain. The organisers were not too happy to hear this, insisting that the trip to the summit was the centrepiece of the programme, but Isaacman repeated that there was a safety principle at stake, and reminded them of previous years' fainting-child debacles. The organisers eventually agreed to settle for a lecture without a summit trip but called a few weeks later to announce, with considerable irritation, that probably as a result of the summit field trip being removed from the agenda *no one* had signed up for that summer's programme! Sometimes being right can make you unpopular, but Isaacman was glad that Peter Forster backed him up completely.

## To Komohana St

Once the decision had been made to have the UKIRT sea-level base in Hilo there were three building options; leasing an existing commercial property and fitting it out with the spaces required, erecting a system built shell and fitting it out on land leased from the state or county or constructing a purpose built building on the University of Hawaii Hilo (UHH) Campus extension. Each of these had its merits in terms of speed, cost, location and long-term planning. Malcolm Longair favoured the last option, though for the team itself it would mean spending a further 18 months in the cramped conditions of Leilani street. According to Kevin Krisciunas, back in 1982–1983 the new office '*Always seemed to be 18 months away*'. He remembers that one time Malcolm Longair mentioned that the plans for the new office were moving along, and it would be '*Ready in 18 months*'. So Kevin asked "*Malcolm, does that mean that next month it will be 17 months away*"? This question was apparently not appreciated.

Construction on the University of Hawaii (UH) Hilo Campus were not part of the remit held by the IfA in Honolulu for development of Mauna Kea so Terry Lee worked with the UH Vice President for Administration Harold Masumoto, the vice chancellor of UH Hilo John Kofel and the UH planning department. The long-term plans for the Hilo site included a new library, a Science Park makai [on the seaward side] of Komohana street and a Technology Park mauka [on the landward side]. There would also be internal roads and bridges linking to the existing campus and (eventually) an entry to the Science Park from Komohana Street.

The Science Park area had been partly marked out in plots and Terry Lee knew the location of the area. Part of it had been roughly cleared for stocking material for the construction of Komohana Street itself and the rest was lava overgrown with scrub. One day he took a pickup truck with a step ladder and one of his crew and he checked the view from various points on those plots which were reasonably accessible. He made a choice of plot (the one now occupied by Gemini) and contacted Harold Matsumoto. A few days later Matsumoto replied that the planning department could offer the adjacent plot, but not the originally chosen one. Lee happily accepted and later received the confirmation in writing only to get a phone call soon after saying that the planning department now wanted him to take the original plot! After some discussion it was agreed that it would be unreasonable to expect UKIRT to chop and change and the arrangement stood. There then ensued some discussion on the size and boundaries of the plot. Terry wanted the entire area from Komohana Street to the edge of the escarpment and argued that the British would probably want a soccer pitch for recreation so they needed plenty of space. Terry describes his thinking thus '*My vision was to place the building close to where the ground falls steeply away to profit from the magnificent views over Hilo*

*Bay, as well as there being visibility between us and the campus and from areas in Hilo. The anticipation also for the longer term that the plans for the development of that part of the campus, would materialise and we would be part of the whole and not planted on the edge and fronting Komohana Street*'.

The UKIRT team engaged a Hilo architectural firm called Oda McCarty to design a building with a combination of offices, laboratories, workshops and a library. The designs were refined in conjunction with Lee and David Beattie with further details specified by a working group of the staff who would be occupying the building. With JCMT development already underway, and a clear need for a JCMT base in Hilo eventually, the design of the entry area was modified so that a second phase could be added to provide additional space when required. The design went out for bids to local firms to construct the facility in the summer of 1984 with the expectation that the building would be complete in about a year.

A building workers strike introduced a six month delay in the programme but, after three years of planning, a ground breaking ceremony for the new facility took place on the 23rd of July 1984. The event was led by Terry Lee and included a record number of speakers including Mayor Matayoshi of Hilo, Eric Becklin, the acting Chancellor of the University of Hawaii, Ralph Miwa, and Councilwoman Helene Hale who had been present at the dedication of various astronomical facilities over many years. At the time the idea that the location of Komohana St might become the kernel of a new technology park was enthusiastically described by the Chancellor, although it would be a decade before much came of this idea. After a blessing was conducted by Father Joseph Priestley the assembled cast of dignitaries and others attacked the ground with their golden shovels and work could begin.

The new building was dedicated in the afternoon of the 9th of August 1985 at an event attended by 200 people. The keynote speakers were Malcolm Longair, Ralph Miwa, and Eugene Tiwanak. The weather was favourable and there were inspiring words from Malcolm Smith and Malcolm Longair, emphasising the commitment of SERC to Hawaii. Malcolm Longair drew attention to the fact that not only was UKIRT already operational on Mauna Kea but that a second telescope, the joint UK/Netherlands JCMT, was already under construction on the mountain. This, he said, emphasised the long-term commitment of the UK to Hawaii. He also pointed out that a second phase of the building, to house the increasingly large numbers of engineers coming out to work on this second telescope had already been approved at the highest level and that this expansion would start soon. For the moment, the millimetre-wave telescope team were occupying the space in Leilani St vacated by the UKIRT staff who had moved into Komohana St.

**Fig. 7.12** Work begins on Phase 2 of the Komohana St building (Photo ROE, origin unknown)

Phase 2 of the new building was completed in the summer of 1986 with its official opening on the 4th of August. For many years the entry was from Komohana Street, a busy main road. Regular users of this entrance remember that this was a bit tricky as there were no turn lanes and it was often a case of pulling into, or more often across, quite fast moving traffic. Originally the County had asked for an exit lane to be built as a condition of planning permission, but the cost of this would have been significant so Terry Lee explained to the County Engineer that the entrance would be temporary and that the staff would not all arrive and leave at the same time, so traffic problems were not likely to arise. Terry remembers that the county officials were very helpful and that in the end the condition was waived since the long-term entry would be from the internal campus road. Of course, at the time there was no internal campus road, nor indeed an entry to the campus!

## ADAM

By the time the VAX computers had arrived at UKIRT a decision had been made to introduce a completely new software architecture to be called ADAM. This effort was led by Dennis Kelly at ROE and, with the intention being to use ADAM on the UK Starlink data reduction network and its potential adoption on other telescopes, it became a multi-observatory project. The involvement of other groups meant that more effort was available to do the work, but it also meant that there were more requirements and overall it made the project harder to manage. Thus the early ADAM days were rather fraught, with the UKIRT team in Hawaii constantly wondering when it was going to be deployed at UKIRT. When it did eventually arrive one of the staff involved in helping to bring it into use was Gillian Wright, who was given the task of helping Ian Gatley to commission the system soon after she joined UKIRT.

Gillian had became an ROE research "fellow" in 1985, soon after completing her PhD at Imperial college. After six months or so in Edinburgh she transferred to Hawaii in 1986. She wanted to work at UKIRT because '*It was the best place to do infrared astronomy*' and because she '*wanted to work with people like Ian Gatley and Tom Geballe who were already established infrared astronomers*'. Although officially in a research post, Gillian found herself heavily committed to support astronomer duties, helping to fill the gaps created when Peredur Williams and Andy Longmore had returned to Edinburgh. Although there was some concern amongst the staff that a research fellow on a fixed term appointment was coming out and doing a job normally done by a permanent staff member, Gillian had "hands-on" experience of infrared instruments and knew how to make them work. Her experience was not limited to instruments, during her PhD she had noticed that in the thermal infrared UKIRT's performance was no better than the much smaller IRFC on Tenerife. She tracked down part of the problem to UKIRT's chopping secondary, which as noted earlier, used a hydraulic system to move the mirror. From time to time the top end of the telescope had got so cold that the hydraulic oil had frozen and a heater had been fitted to prevent the top end from freezing up. This had the effect of slowly changing the temperature of the top-end so that the thermal background drifted throughout the night, reducing the sensitivity of the instruments operating in this wavelength region.

On ADAM Gillian says '*The users initially hated it, because there was no system in those days for proper version control and there would keep being new versions that we'd just get going by 6 pm (the start of observing) and then we would find later in the night that it still had assorted bugs and difficulties which were imposed on the users, as we couldn't easily go backwards. It was a major issue to reinstall the older software and the switch was made in a rather brutal fashion. Ian Gatley had little tolerance for user moaning – he was totally focussed on the advantages it would bring and considered that the approach would mean it just had to be made to work. (A short sharp pain for a long term gain sort of idea)*'. Gatley was an early and strong advocate of the idea that software was the key to revolutionising infrared astronomy. Being able to nod the telescope without what he called "a human Dickey switch", often otherwise known as a PhD student, was a major achievement of ADAM, as were the maps made by scanning the sky with single element photometers using the system. According to Gillian it '*would have been unthinkable without ADAM to make that scale of a map with a UKT instrument and it was unique to UKIRT for a while*'.

# Chapter 8
# IRCAM: The Beginning of the Array Revolution

By about 1981 the UKIRT Users Committee was considering options for future instruments. Infrared array cameras were amongst the possibilities although, ironically as it turned out, a camera for the 1–5 μm regime was not the top priority at the time. Nonetheless, Malcolm Smith (then head of instrumentation at ROE) and Malcolm Longair decided that the ROE needed to find the infrared equivalent of the optical CCD to replace the single element detectors then in use in UKIRT's photometers. Fortunately they already had someone for the job at the observatory. His name was Ian McLean and he had arrived in November 1979 after a period as a post-doctoral research associate in Tucson where he had been working on solid state detectors for optical astronomy. Ian had come to Edinburgh on a temporary contract to work with Ray Wolstencroft to build a CCD-based imaging spectropolarimeter before joining the ROE staff two years later.

Since McLean was already working on CCDs, the task of developing an infrared imager was assigned to him. His first step was to do a survey of what type and size of infrared detectors were available. He started in the UK because ROE technical manager Donald Pettie had contacts at Mullard in Southampton, where early work was being done on Mercury-Cadmium-Telluride (HgCdTe) detectors. These devices turned out to be for longer wavelengths than were wanted for astronomy and McLean was told that he would be better looking in the USA for near-infrared arrays. Such detectors did exist, but the technology was for military use and only a small number of early examples had slipped out into the astronomical community. So at first sight the prospects looked bleak; it seemed that if you didn't have some contacts in the US military it was impossible to get one of these devices and, even if you did, they were not all that good for astronomy.

He was advised to visit John Rode at the Rockwell Science Center in Thousand Oaks, California and so, in late 1982, he set off for a tour of the USA. Rode took McLean into his laboratory late one evening; long after most of the staff had left, and showed him some of the devices they were developing. On the same trip McLean also visited the nearby Santa Barbara Research Corporation (SBRC), which had provided single element InSb detectors to UKIRT and were known to

 
J.K. Davies, *The Life Story of an Infrared Telescope*, Springer Praxis Books,
DOI 10.1007/978-3-319-23579-0_8

be working with InSb arrays. Here he met astronomer Alan Hoffman who gave him a tour of the SBRC facility and who says he had *'Always wanted to use the new infrared arrays for astronomy from the time I started at SBRC in 1979'*. Ian also visited Rochester University in New York State whose astronomers had a 32 by 32 infrared array but, as he put it, '*Were not saying where they had got it from'*. Aware of the sensitivity of probing too hard McLean said to his host Judy Pipher, '*OK, I know you can't tell me where it came from but can you tell whoever gave it to you that we, the UKIRT, are interested in such a device*'.

Pipher did as she was asked and it turned out that the device had come from Hoffman at SBRC who had, as he puts it, "borrowed" a 32 × 32 array that had been manufactured using internal funds and hand-carried it to the University of Rochester. Initially Hoffman says he was in trouble for unauthorized use of the array but, with the help of Carol Oania of the SBRC marketing department, he was able to convince his management that this was a new business opportunity. According to Hoffman *'I wanted to do it for science, but Carol wanted it for the money it would bring to SBRC. Her approach was a huge advantage when dealing with management. She spoke their language and made it possible for me to design and build the arrays for astronomy'*.

Thus Oania and Hoffman persuaded SBRC to start a commercial programme to develop IR detectors specifically for astronomy. At least then, they argued, we can do some (unclassified) work which we can talk about to our friends and families. Working with the UK was unusual for a US contractor, but since the UKIRT telescope, the intended destination of the device, was in Hawaii that helped smooth things over. Malcolm Longair's support and energy was also an important factor behind the deal, with Longair promising that ROE would provide whatever resources necessary to make the camera a success. The UKIRT team were not alone in their interest in such a detector, Al Fowler at the US National Observatory on Kitt Peak (NOAO) was already talking to Hoffman about astronomical arrays. Eventually both ROE and NOAO would play a role in the development of the arrays, with information flowing back and forth with the manufacturer in a synergistic way when problems arose.

So in 1984 SBRC began to build an InSb array for astronomy, a device designed to work in a low-background environment and without the need for real time readout. These arrays were not CCDs, they were hybrids with an infrared detecting layer on the top and a read-out device underneath. The two layers were joined by tiny (a few microns high) individual columns of indium which provided individual electrical contacts between the pixels in the infrared layer to those in the readout device underneath it. The final negotiations with SBRC took place in Hawaii in March 1984 and included a visit to the summit of Mauna Kea to see the facilities there. The SBRC management, led by Dr. Curry, a vice president of the company, was convinced by the visit, and perhaps encouraged by the sight of nearby Mauna Loa erupting, a contract was signed. A few months later, McLean and Tim Chuter, who would be the engineering manager for the camera, visited SBRC to go into the technical details. Here they got an unpleasant surprise when the SBRC team told them that the original low-noise design they had been discussing could not in practice be made and the alternative design had a much higher read noise, perhaps a factor of 10 worse. In the end, the ROE team decided things might not be so bad, and decided to go ahead anyway.

The first meeting of the IRCAM project team, which would eventually grow to about 11 full-time staff including McLean as project scientist, Tim Chuter and software engineer Dennis Kelly, was held on the 2nd of July 1984. The project quickly developed a design that used external mirrors on a platform to produce a parallel beam going into the cryostat window and thus kept open the option of putting a Fabry-Perot etalon directly in front of the window. The cryostat was a box about 50 cm square with two tanks, the bottom one containing liquid nitrogen and the top one liquid helium. This dual-cryogen arrangement was chosen since the expected operating temperature of the detector was about midway between the 4 K of liquid helium and the 77 K of liquid nitrogen. By balancing the two coolants against each other it was possible to reach the required temperature range and adjust it as required. The operating temperature was controlled by a gas switch, a partly evacuated copper tube containing helium gas and a small heater. When in contact with helium in the dewar the gas in the switch would attempt to liquefy, but some of it could be re-evaporated using the heater. The variable amount of gas changed the thermal conductivity of the switch and so the equilibrium temperature of the detector could be controlled.

Inside the cryostat there were just three mechanisms, one for focus and two filter wheels, all driven by external motors via shafts going through the cryostat walls. The early thinking about the design had included an internal lens turret that would have carried several alternative lenses to give different pixel scales with correspondingly different fields of view. However, in order to speed development and get IRCAM to the telescope as quickly as possible this option was removed and the eventual design allowed for only one lens to be fitted at a time. Three alternative lenses were available, giving pixel scales of 2.4, 1.2 and 0.6 arcseconds per pixel, but changing the plate-scale, and hence the field-of-view, would involve opening up the instrument, replacing the lens and then putting the camera back together.

Final assembly started in 1985 and about a year later IRCAM was complete and under test at ROE. The detector was a 58 by 62 array, giving 3596 pixels, which is ludicrously small by today's standards but, as McLean put it "was 3595 pixels more than was available before". A close look at the outer edge of the array revealed the microscopic logo "tank breaker", clear evidence of the device's military heritage. On the 26th of March 1986 the team had first light of IRCAM in the ROE laboratory. The result was described by Ian McLean as '*rather like looking out of a window at a night time scene through vertical venetian blinds- some open, some closed*'. The moment of triumph was only spoiled by the fact that all of the "blinds" should have been open. This problem was soon traced to a short circuit caused by some solder and was quickly fixed. A much improved image, showing a simple logo spelling out "ircam", was obtained on the 9th of April. Further testing cycles followed to investigate the effects of detector temperature on performance (a capability which stemmed from the installation of the gas switch and which enabled ROE to feed back performance information to SBRC as part of the synergistic supplier-customer relationship) and these showed that the optimum operating temperature of the array was about 35 K. These first tests were done with an engineering grade array and after few months this was replaced by a better quality array more suited to doing scientific observations.

Ian McLean moved to Hawaii in the summer of 1986 to get settled in and then returned to Edinburgh to prepare IRCAM for its shipping to Hawaii. On the 26th of September, 12 crates containing IRCAM and its support equipment were shipped to Hawaii. Colin Aspin, who had been writing software for the system, would also transfer from ROE to Hawaii at that time. McLean and Aspin were soon joined by Edinburgh University PhD students Mark McCaughrean and John Rayner who came out for observing runs. Gillian Wright acted as the local scientist. On arrival IRCAM was set up in the laboratory at the JAC in Hilo for complete end-to-end testing before taking it to the summit and mounting it on UKIRT.

The astronomical "first light" for IRCAM was on Tuesday the 21st of October 1986, but it was not at night. Given the track record of failed or marginal IR cameras in the past, Malcolm Smith, by now UKIRT director, would only give them morning twilight for this early testing; he did not want to risk wasting good observing time. So that morning, as Orion sank into the western twilight, McLean and Aspin, joined by Gillian Wright and Dolores Walther, took a K band (2.2 μm) image of the Orion nebula. It looked good and the two principals smiled at each other. McLean described the fact that they could take such a picture in what was by then broad daylight *'seemed to stun everyone into silence'*. Then he remembers Gillian Wright saying *'I don't believe you two, this is a momentous occasion and you should be jumping up and down and congratulating each other'*. According to McLean the reason that they were not so excited was that they had taken over 1000 test images in the laboratory and they knew that IRCAM worked. To them, taking a picture of something you could see with the naked eye wasn't what it was all about. The next object they observed was the BN object, a true infrared source with no visible counterpart. On reflection though he says *'Gillian was right, it was a momentous occasion so we did shake hands and clap a little bit'*.

Within a few days some adjustments had been made and better images were obtained, albeit still in the daytime, and soon after the first night time observations were made, allowing the camera to start observing fainter objects such as galaxies. It was a highly successful start to the IRCAM commissioning and on the 27th Malcolm Smith sent a message to ROE congratulating everyone there on the hard work and professionalism that had gone into the project and remarking that it had shown a level of excellence that would be *'Acknowledged throughout the international IR community'*. He also said, more informally that *'Carpet wear in our building in Hilo has reduced to zero as people's feet do not seem to be touching the ground'*.

However, ground-breaking though they were, the original IRCAM images were not all that good because the detector had a diagonal crack across it. Early in 1987 SBRC sent a new array that was much better with fewer bad pixels and lower read noise. Soon images were being taken of all sorts of objects; Jupiter, Halley's comet, planetary nebulae, supernova remnants and still fainter galaxies. Some images were taken through narrow-band filters with sub-arcsecond pixels, images that many people had thought were never going to be possible. The astronomers also learned how to mosaic together images to create pictures covering larger areas, presaging the days when much larger arrays would become available. One famous early mosaic was that of the Orion nebula. The images, in the K filter, were taken in early December 1986 and Mark McCaughrean took a few days to mosaic all the frames together (it took about 100 hours to do). McLean described it as *'A nice Christmas present'*. The image was presented as a late poster at the January meeting of the American Astronomical Society and Ian recalls that *'I think we knew by then that IR astronomy had changed forever'*.

Mark McCaughrean recalls an incident at the meeting which to him showed the paradigm change that was taking place. He said in a meeting in 2009 that *'A senior IR astronomer wandered past the poster and gave the image the once over. He didn't seem particularly impressed, to be honest, but just before departing, he asked a question in a rather derogatory tone which was so out of left-field, that I was completely dumbstruck. The question was "Why are some of the stars bigger than the others?"'* The question illustrated the mental switch between using old style photometers like UKT9 to map extended sources (for example gas clouds) by scanning the telescope across them and imaging them by pointing a camera and integrating the signal on a fixed position to build up signal to noise. True the photometers and their scanning software was well developed and the read-noise in the early infrared arrays was very high, but the noise would integrate down during the exposures and better arrays would surely come. There was also an issue of pixel size. Photometers could be operated with large apertures to collect faint light over a large area and the relatively tiny pixels in IRCAM would not have that light gathering power and so would miss very diffuse emission. However, not all such emission was expected to be diffuse and the better spatial resolution of the array cameras would soon confirm that much of it occurred in bright knots, which were smeared out, and so not detected, by mapping with larger apertures. In the end, it was the proponents of IR cameras who would win the day, but not everyone would concede this at the time.

That astronomy had changed was confirmed at the "IR Astronomy with Arrays" meeting held in Hilo in March 1987. Although intended to be an informal workshop the event attracted over 200 people from ten countries and the presentations from the ROE staff about IRCAM were an outstanding success, causing some American astronomers to remark that IRCAM had *"stolen the show"*. Indeed Wayne van Citters, speaking about the impact of IR arrays for astronomy, said *"I don't think there's any doubt that we are on the verge of a promised land in optical and infrared astronomy"* and the director of the IfA Don Hall added *"Things that seemed in the future and beyond my grasp for so long are clearly here."* People at the meeting didn't yet know what was going to come in the future but they did know that they were present at a time of change. John Rayner attributes the success of IRCAM compared with its contemporary rivals to *'The emphasis Ian gave to the software development and in particular data reduction and display'* and Bob Joseph echoes this thought *saying 'Infrared astronomers had virtually no experience with array data, and if science were going to come out from IRCAM the software to reduce and process images was just as important as the hardware'.* Ian Robson wrote in 2013 that *'This was a mammoth effort that should not be underestimated. The continuous push by Ian Gatley to have the software developed so that science could be done from mosaiced images, where the background sensitivity and edge effects were removed, paved the way to opening up a whole new area of infrared astronomy'.*

**Fig. 8.1** Ian McLean with some early IRCAM images (Photo ROE)

Despite, or perhaps because of, his success with IRCAM Ian McLean did not remain in Hawaii for long and by 1989 he had left to take a position at the University of California, Los Angeles where he continued to develop infrared instruments. His position as IRCAM scientist at UKIRT was taken by Mark Casali who had joined ROE after completing a physics PhD at Melbourne. An Australian, born of Italian parents but working for the British in the USA, Mark quickly became an expert on UK work permits and US visas, but managed to make it to Hawaii, via brief stop-over in Edinburgh in April 1987. Colin Aspin took on the related responsibility of managing UKIRT's polarimetry modules, which were often operated with IRCAM as the detection system.

Around the same time a second IRCAM was delivered from Edinburgh. Having two cameras meant that each could be permanently set up for a particular field-of-view, removing the need for the potentially risky process of repeatedly warming up the camera, opening and closing it and then cooling it down again every time a plate-scale change was required. The second IRCAM had the hybrid liquid nitrogen/liquid helium cooling system of IRCAM 1 replaced by a two stage mechanical closed cycle cooler in late 1990. As well as saving the cost of the liquid cryogens this removed the need for daily human intervention topping up the coolants, saving time for the technical staff during normal working days and allowing the instrument to be left unattended over weekends and holidays. It also reduced the risk of an unplanned warm up if, due to bad weather, the telescope building was inaccessible for a few days. IRCAM 1 received the same modification the following year.

**Fig. 8.2** John Rayner does a helium top-up of IRCAM before the closed cycle coolers were fitted (Photo ROE)

## Slipping and Sliding

An early user of IRCAM was Karl Glazebrook who at the time was doing a PhD in Edinburgh and had an interesting night on the 28th of March 1988, during what was only his second observing trip to Mauna Kea. There had been quite a bit of snow in the preceding few days and although the road had been cleared, a thunderstorm in the afternoon had deposited a lot of hail on the freshly cleared surface. Although the weather did not look promising for observing, Joel Aycock needed to drive up the summit to top-up the cryogen tanks in some of the UKIRT instruments so Karl and his supervisor John Peacock went along with him to see what the conditions were like. Karl recalls that quite a nasty thunderstorm was developing and, just as they were approaching the summit, they came across a wrecked Chevrolet Suburban car belonging to the University of Hawaii. Although it was the right way up they could see from the trail on the mountainside that it had rolled down from a switch-back further up. Joel describes it as looking like *'a crushed beer can, with the roof smashed down to the door panels and severe damage on every side; front, back, sides, top and bottom'*. The three men approached the wreck with trepidation, half expecting that they would find bodies of friends trapped inside.

Fortunately the vehicle was empty, with no signs of blood or even of footprints in the snow, so they carried on along the road to see what was going on higher up. They soon found where the vehicle had gone off the side (it was the last switch-back before the summit ridge) and saw UH staff member Frank Cheigh and a group of Japanese astronomers searching for their bags and papers which had been scattered on the hillside as the vehicle rolled downhill. They stopped to assist and suddenly found their own vehicle sliding backwards on black ice. Fortunately Joel managed to stop the car by changing the direction of the slide to the safe side of the road and they all got out. About the same time Charlie Telesco, an American astronomer en-route to the IRFT telescope, arrived.

**Fig. 8.3** The crashed University of Hawaii vehicle at the end of the trail it made as it rolled down the mountainside (Photo Karl Glazebrook)

Interestingly, although most of those involved describe their recollections as vivid, it is difficult to reconstruct the precise series of events. Karl remembers that while they were talking yet another telescope vehicle arrived and did exactly the same thing; despite desperate signals to keep moving the car stopped to help and started sliding too. John Peacock thinks that this last car stopped uphill of the UKIRT vehicle before it started sliding back, and that the UKIRT people helped to stop it sliding. However he cannot remember if this was Charlie Telesco's car. In any case Charlie Telesco recalls that *'when we started to go on our way, our vehicle started to slide sideways towards the precipice. That was* **very** *scary, and it took several of us pushing on the outside of our vehicle to keep if from going over the side. What added to the scariness was that several of us were pushing from the cliff-side of the vehicle, which was slowly pushing us toward the edge. We did overcome that, but it was touch and go and very dangerous'*. Possibly, all the accounts are correct; the situation was clearly stressful and confused and it all happened 25 years ago so today none-one is quite sure exactly how many vehicles were involved.

In the event Karl, John and Joel continued to the UKIRT building, but it was clear that the storm was getting worse and eventually the decision was made to abandon the mountain for the night. The UKIRT car was equipped with tyre chains and to minimise the risk to life it was decided to walk the mile or so down to JCMT, except for one person who would drive each vehicle. By now they were at 14,000 ft, in the middle of a blizzard with sheet lighting flashing all around the mountain and lots of thunder and cloud. To Karl *'It felt like a scene from Ragnarok. It was a very scary walk, the electrical energy in the air all around us was palpable'*. Everyone's hair was standing on end due to static and the metal poles along the side of the road were audibly crackling; they were actually discharging in to the air. Karl *'Had the horrible feeling we would all be zapped by lightning at any minute'*. Luckily they were not electrocuted and soon rendezvoused with the vehicles and drove safely down the mountain.

Safely back at Hale Pohaku, Joel spoke to Frank Cheigh, who had been driving the Suburban. Joel says *'He was still white as a sheet, in mental shock after the accident. The story I got from him was that he had gone to the summit for instrument fills, just as we had. With him were three Japanese astronomers, only one of which spoke a little broken English. After finishing up, they were headed back down, but when they braked for the first turn below UKIRT, their truck started sliding slowly. It slid to the right, and the front end caught in the snow piled up by the snow plows, on the down slope edge of the road. As the front end caught and stopped, the truck slowly rotated around that pivot point, until it was facing backward and continuing to slide down slope. At that point, Frank and his passengers exited the slowly moving vehicle, and watched as it hit and broke through the snow bank again (backwards this time), and tumbled down the hill. They ended up hiking back up to CFHT, and got a ride with that crew back to HP. Frank Cheigh, who had worked on the early development of the UH88 starting in the late 60s, and had worked on the summit of Mauna Kea for over 25 years at that point, quit his job just a few weeks later, and I never saw him on the summit again... I don't think he ever got over his experience on the summit that day'*.

Joel believes that the men returned to the summit later in evening to recover their papers, so it is possible that when he, Karl, John and Charlie encountered the group on the roadside it was a few hours after the actual accident. The whole incident is reminiscent of the near miss of the UKIRT car carrying Terry Purkins and others during the construction phase of the telescope and shows just how close UKIRT had come to disaster some years earlier.

## IRCAM Upgrades

The IRCAM design had allowed several potential upgrade paths and these were implemented in later years. The first such change was the development of a new non-destructive readout scheme for the array in mid-1990. This reduced the amount of readout noise and allowed IRCAM to operate more efficiently at longer (3–5 μm) wavelengths where many rapid readouts of the array were needed to avoid saturation of the detector by the strong background emission from the night sky. In 1991 a two times magnifier that screwed onto the outside of the cryostat was provided, offering the option of operating IRCAM with a pixel scale of 0.31 arcseconds per pixel. Further upgrades of the basic IRCAM would take place a few years later but these would be the responsibility of another support astronomer, for Mark Casali left the JAC for Edinburgh in March 1992.

In 1994 IRCAM's small, and by now quite outmoded, 58 × 62 array was replaced with a 256 × 256 InSb detector and fitted with a new array controller called ALICE, the Array Limited Infrared Control Electronics (the "E" was soon changed to Environment since ALICE was more than just electronics, a lot of software was involved). The upgrade was led by Phil Puxley and involved rewiring the second IRCAM cryostat and fitting a new lens assembly. The upgraded instrument, now known as IRCAM3, was commissioned on the telescope in April 1994 and performed extremely well from the start, with few cosmetic defects on the array and generally high throughput. The upgrade restored IRCAM to a state-of-the-art instrument competitive with similar cameras elsewhere in the world.

## Dome Repairs

In August and September 1988 the telescope was taken out of service for a major repair and refurbishment of the dome. New electronics were installed, the base was refurbished and the existing bogies were replaced by wheels that ran on a rail around the top of the dome wall. This made dome rotation both quieter and more reliable. At the same time various software and cabling improvements were made and the opportunity was taken to refit the control room to provide a better working environment for the observers by providing new furniture, new lighting and fresh paint. The control room stereo system was also upgraded to include a CD player.

**Fig. 8.4** The new look UKIRT control room (Photo Ian Robson)

## International Agreements

During 1989 an informal agreement was reached for Canada to trade unused JCMT shifts, to which they were entitled as a partner in JCMT, for UKIRT time. A similar deal existed with the Netherlands under which unused Dutch observing time at La Palma and on the JCMT time could be traded for UKIRT time. However, in later years it seemed that the scheme was not being used and it soon lapsed.

# Chapter 9
# Cooled Grating Spectrographs

## CGS2 Upgrade

CGS2 had been UKIRT's near infrared spectrometer since its arrival in 1983 and, although it had been very successful, its sensitivity was limited. So in 1988 a plan was put in place to upgrade the instrument to prolong its useful life until a new spectrometer became available. The upgrade replaced the existing detector amplifiers with new devices able to integrate the signal on the detector and so reduce the effect of electronic noise, which occurred only at the beginning and end of each data taking cycle. It also added a Hewlett Packard real-time computer to the detector control system, which meant the readout algorithm was much more flexible and CGS2 was one of the first instruments in the world to use non-destructive reads to reduce the readout noise (albeit with only seven channels). The improvements to the instrument were commissioned over the 16th to 18th of February 1989.



J.K. Davies, *The Life Story of an Infrared Telescope*, Springer Praxis Books,
DOI 10.1007/978-3-319-23579-0_9

Unfortunately, the improved sensitivity of CGS2 was compromised because one of its seven detectors had stopped working and another, number 5, was unstable. Thermal issues also meant that the background inside the cryostat was both high and variable, degrading sensitivity. After some careful planning and preparation a team of "surgeons" led by Sidney Arakaki opened the cryostat in August 1989 and replaced the detector and amplifier package and worked to improve the thermal coupling between the cold radiation shield and the internal optics. The repair was a complete success; all the detectors were restored to operation and the internal background was reduced by an order of magnitude. Although there were still some problems with detector spikes, which had always been present and which also affected the UKT instruments, it was decided that the time and effort required to improve the software in an attempt to ameliorate this problem was not justified in view of the imminent arrival of a new spectrograph, CGS4, which would soon render CGS2 obsolete.

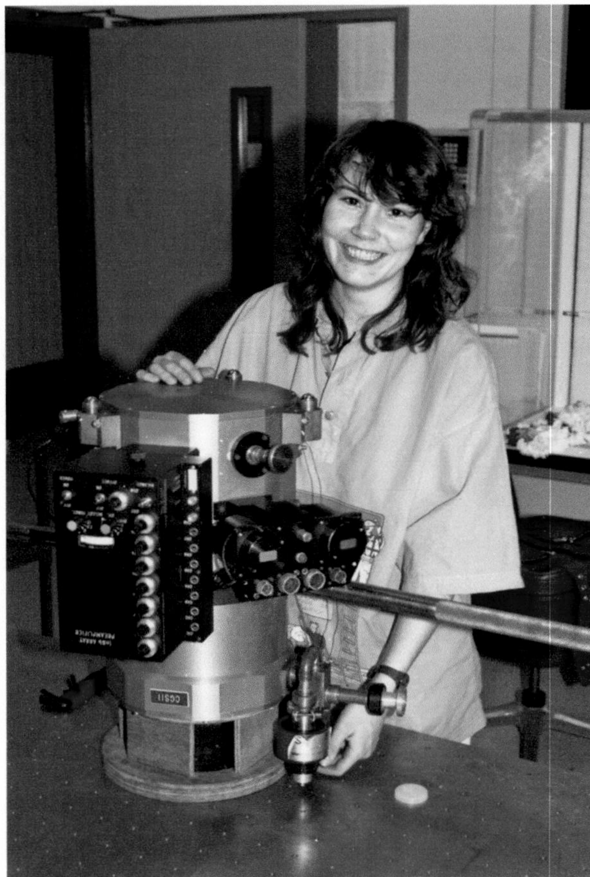

**Fig. 9.1** Graduate engineer Rosemary Glendinning (aka Chapman) and CGS2 during the upgrade (Photo Steven Beard)

# CGS3

While CGS2 operated in the near infrared, between 1 and 5 μm, UKIRT also needed the ability to do spectroscopy at longer wavelengths, in the thermal infrared windows around 10 and 20 μm. This region had been explored by, amongst others, a series of successful mid-infrared spectrometers built and operated at telescopes around the world by Dave Aitken and Pat Roche from University College London (UCL). However, when Dave Aitken moved to Australia, he took his instrument with him and from about 1982 onwards UK astronomers had no national access to a common-user mid-infrared spectrometer. It was to fill this gap that the third in the series of UKIRT cooled grating spectrometers, CGS3, was designed and built by a group at UCL led by Bill Towlson. Essentially an improved version of the Aitken spectrometer, CGS3 used a linear array of 32 photo-conductive Si:As detectors and included three diffraction gratings. The gratings allowed CGS3 to cover the entire 8–14 μm "N" window or the 20 μm "Q" window in single observations, with a higher resolution option available for the N window if required. The instrument was built inside a conventional cylindrical cryostat with liquid helium cooling the detectors to their operating temperature of a few Kelvin. Like IRCAM, it was designed to be mounted on a platform attached to the instrument support unit of the telescope and to receive the telescope beam via a pair of adjustable alignment and collimating mirrors.

The instrument was taken to the Royal Observatory Edinburgh for an extensive series of pre-delivery testing and software development by two ex-UCL scientists, John Lightfoot and Alistair Glasse. These tests, in 1989 and 1990, integrated CGS3's on-board control computer into UKIRT's instrument control environment. They were a forerunner of the pre-delivery integration and testing which later became a key stage in all ROE instrument programmes. The first observations were made at the telescope in July 1990 by a team including Bill Towlson, electronics engineer Martin Palmer, Chris Skinner, student Teresa Hunt from UCL plus Alistair Glasse and John Lightfoot from ROE. A spectrum of the Red Giant star MX Boötis was taken on the 5th of July and revealed the spectral signature of silicate dust grains. In spite of initial teething troubles, dealt with by the technical staff at JAC, by May of 1991 the operation of CGS3 was much improved and it settled down into regular use.

In common with other mid-infrared instruments, CGS3 never used up a large fraction of the telescope's observing time. This lack of popularity with many research groups was primarily due to its operating wavelength range; 10 and 20 μm observing is notoriously difficult since it is a regime where the atmosphere is very bright and the observing windows, especially in the 20 μm, or Q, band can be degraded or even blocked completely by atmospheric water vapour. Secondly, the mounting platform which was required to attach CGS3 to the telescope could only be fitted to the North or South ports, and these were usually occupied by the popular IRCAM and CGS4 (q.v.) instruments. Accordingly CGS3 could only be mounted when one of these other instruments had been removed. Despite these difficulties a small but hardy group of dedicated users persisted to use it for various science programmes, for example in the study of the dusty environments of late type stars and defining infrared calibration schemes. CGS3 would continue in service until the early 2000s when it was superseded by the much more capable Michelle imager/ spectrometer.

**Fig. 9.2** Bill Towlson with CGS3 on UKIRT (Photo Gillian Wright)

## CGS4

The arrival of the first infrared arrays and their success in IRCAM made the next step obvious, similar detectors should be installed in a near-IR spectrograph. While IRCAM plus its narrow-band filters or Fabry-Perot etalons could be used for spatially resolved imaging over a narrow range of spectral lines, a 2 dimensional spectrograph would offer much greater multiplexing in wavelength and provide a complementary facility to IRCAM. Such an instrument would represent a major improvement over CGS2, with its 1 by 7 pixel array and over the CVFs in UKT6 or

UKT9. Using an array offered better spatial and spectroscopic resolution in an instrument that could both take data along a slit, rather than through a circular aperture, and which could cover an entire 0.4 μm atmospheric window in one setting. It would also open up the possibility of using an echelle grating for high resolution spectroscopy.

The initial science requirements were handed down from Richard Wade, who was then Head of the UKIRT unit at ROE, to Matt Mountain who would be the project scientist. Matt had joined the ROE in 1985 after nine years at Imperial College London doing both research and instrument building projects. The initial scheme was that the new instrument should fit inside a standard, CGS2 sized, dewar, weigh less than 100 kg so it could be mounted on the existing instrument support unit, cost less than £500,000 and be delivered in four years. However, in the end the laws of optics and physics got in the way of this ideal. To meet the scientific requirements CGS4, as the instrument became known, finished up being a lot bigger than a standard dewar and weighed more like 350 kg. Since CGS4 was also being designed at a time when IR array technology was advancing rapidly and much larger InSb and HgCdTe arrays were already on the horizon, an upgrade path for these new detectors would need to be built in from the beginning. This added complexity to the design (indeed early in the programme the possibility of building two instruments, one optimised for 1–2.5 μm work with a HgCdTe array and another for 2.5–5 μm work with an InSb array was considered). Ultimately only one CGS4 would be built, and it would cost nearer to three million pounds, but as Bob Joseph said *'It does cost money to do things right'*.

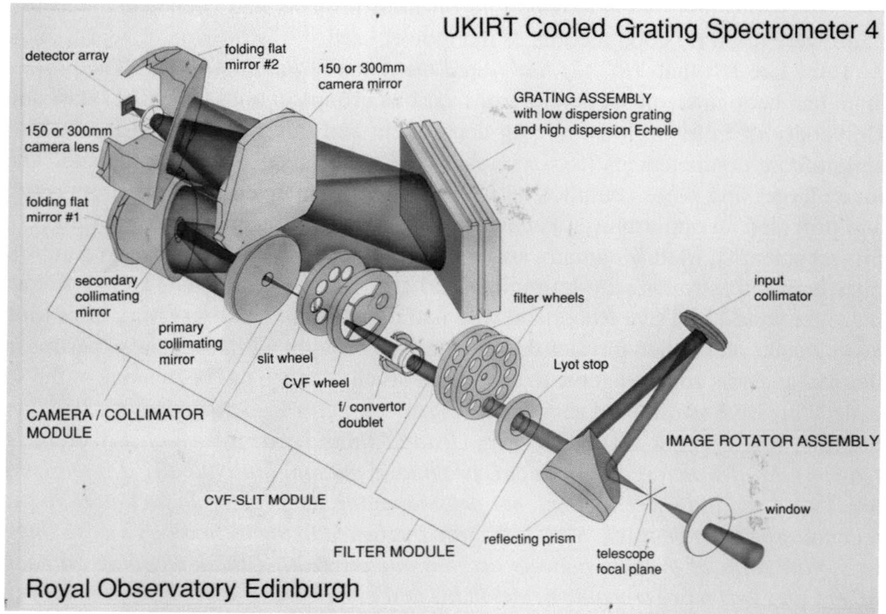


**Fig. 9.3** CGS4 exploded diagram (Photo ROE)

CGS4 represented the culmination of the gradual change in the way infrared instruments were built at ROE. As noted earlier the arrival of Malcolm Longair as Director, and the appointment of Donald Pettie as Chief Engineer, had led to the evolution to a more project based culture. Early UKIRT instruments (the UKT series of photometers for example) had been built by a partnership of one scientist and one or two engineers. Instruments like CGS1 and CGS2 were more complicated and involved a small project team who supported (for CGS1 and 2) Richard Wade and Dave Robertson as Project Scientist and Engineer respectively. The new infrared array instruments such as IRCAM, led by Ian Mclean, took this approach further, working much more openly with regular project meetings of the technical staff and the involvement of wider observatory management.

CGS4 had to take the next step. Following his return to Edinburgh as Head of the ROE Technology Unit, Terry Lee, along with Donald Pettie and Richard Wade had been thinking about a number of management and technical issues with projects of this scale and the way ROE was approaching the building of its increasingly more complicated and expensive instruments. Typically a scientist might gain project funding and then expect to get appropriate staffing from the Technology Unit or from elsewhere in the observatory. But as projects got more complicated, and gained greater external visibility through committees such as the UKIRT Users Committee, much more attention needed to be paid to the selection of projects, resource allocation, engineering analysis and the quality and professional development of engineers and technicians. Technical factors needed to be considered carefully since mistakes became much more expensive, for example thermal leaks were not always thoroughly analysed early enough in a project to make good decisions and, while expenditure on most projects had been accounted for, there were often no good records of manpower used.

Terry Lee felt that *'IRCAM had tested the limits of our capability'*. The project team had been made up of people from various groups in both the SERC staff and University of Edinburgh astronomy department and this had exposed the lack of certain core competencies (for example in software engineering) on the site. The much larger and more complicated CGS4 project demanded a different approach and provided an opportunity to address many of these concerns. There would be a project scientist, Matt Mountain, and project manager who would have to compromise between astronomical desirability and engineering reality. The CGS4 project manager would be Dave Robertson who had returned to the observatory after time away to take an applied physics degree and who brought with him new expertise in thermal analysis and cryogenic techniques. Although much has been made of the so called "creative tension" of such an arrangement, Dave Robertson is not a believer that this was a major factor. He says *'I don't think I was aware of any creative tensions. Matt was a quite practical, pragmatic person, but typically a scientist. I was interested in the technology, not just managing the project, in fact more in the technology than managing. Matt was quite interested in the technology too, so there was quite a lot of overlap between us and our interests. I think stepping on each others toes was more a problem, me on his and him on mine. I would tend to view it that way, more than the other way around. We did have blazing rows from time to time, usually about something trivial, but we got around those and got on with building the instrument'*. One of these trivial matters was the colour of the CGS4 cryostat. According to Dave *'It is popular myth that the colour was picked by David Laird and myself because we were Hearts supporters but anyone who knows Hearts or has been to see them perform will know the CGS4 colour is nothing like the maroon they play in. In fact the colour was picked to buck the trend of producing instruments in "UKIRT Blue" as the IRCAMs were. This didn't stop Matt Mountain and I having a blazing row about it – I think I literally threw a management book at him on that occasion. It was Managing Professionals in Innovative Organisations I believe'*.

As the Head of the Technology Unit Terry Lee puts it this way *'The management structure we wanted was one where the project scientist and project manager/engineer had equal status representing different inputs and a shared responsibility to deliver. Projects with a dominant scientist can have a tendency to have deficient technical performance, inadequate advance planning and poor resource control. Projects without adequate input from a project scientist rarely end up with a satisfactory outcome. Whatever the management theory and jargon this worked well'*.

Dave and Matt would lead a team effort involving optical, mechanical, electrical and software engineers using a systems engineering approach to bring together all the disciplines required to complete the project. Malcolm Stewart, who had returned from Hawaii at the end of 1982 and had just completed a part-time PhD while working in the UKIRT Home End at ROE was now available to lead a software group. There was an external project scientist, initially Mike Selby and later Bob Carswell from the University of Cambridge and, in collaboration with the University of Edinburgh, PhD student Suzie Ramsay was embedded in the project team. Suzie had come from a first degree at Glasgow and was determined, despite attempts to persuade her otherwise, to do a project that involved working on instrumentation and not just doing pure research. She would spend much of her PhD doing optical alignment and testing of CGS4 in the ROE laboratories. So, in many ways, the CGS4 project became a forerunner of all the large instruments developed for 4 and 8 m telescopes since.

The size and scope of the project required innovations in almost all areas of the design and in its manufacturing. CGS4 was much bigger and more complex than anything that the ROE had done before so there were numerous concerns about basic issues such as mechanical flexure and cooling. To deal with these the instrument team developed an innovative optical design that required aspheric, and hence diamond-machined metal, mirrors all to be mounted in an aluminium structure to avoid any thermal distortions when cooling the instrument down to cryogenic temperatures. An additional advantage of using diamond-machined optics was that it enabled precision reference surfaces to be fabricated and these would greatly facilitate assembly to the tight optical tolerances required without needing the traditional kinematic mounts that are typically used in high precision optical instruments. Furthermore, this approach of all-metal optics mounted like plates rather than precision optics structures ensured predictable thermal performance both in terms of temperature and distortion within the cryogenic support structure. To provide confidence in the design, and then build it to the required tolerances, CGS4 used a host of other then new technologies; the Code-V ray tracing systems for optical design, Finite Element Analysis for flexure modeling, detailed thermal modeling of the cryostat and CNC machining for component manufacture.

Mechanical stability was a big issue and rather than using a traditional truss structure, CGS4 used a monocoque casting in which the skin of the internal boxes provided the required mechanical stability. To move the various gratings, focus and other mechanisms required eight motors and this raised concerns about how to do this without compromising the thermal design by incorporating too many pass-throughs from outside the cryostat. The solution was to use, for the first time, stepper motors that could be operated at cryogenic temperatures, a decision which eliminated the need for pass-throughs and made the instrument more compact. CGS4 also used space-qualified bearings inside the cryostat to maximize reliability. Traditional instruments used liquid cryogens for cooling but later models of IRCAM had used mechanical cryocoolers and CGS4 built on that experience, using closed-cycle cryocoolers for normal operations while adding a capability to use liquid nitrogen for rapid pre-cooling whenever the instrument had been warmed up for engineering work.

At the heart of the instrument was of course the detector array. The early $58 \times 62$ arrays were fairly low-noise for the time, but in spectroscopy the instrument disperses the background light from the sky across many pixels so electronic noise generated when reading out the detector can easily dominate the weak signals of interest. Accordingly, CGS4 used various techniques like non-destructive reads of the detector while the signal ramped up and then fitted the resulting data to reduce the impact of detector noise. Another innovation was the idea of detector translation, in which the detector was moved slightly during the observation to create several sub-spectra, which could be interleaved to fully sample the 0.4 μm wide window and mitigate against any bad pixels in a single observing sequence.

CGS4 was more than just an instrument, it might be better described as an observing system since it was equipped from the start with a sophisticated software package. This allowed astronomers to define sequences that could operate the instrument and offset the telescope as required. The control system could also chop the telescope when observing in the thermal infrared (wavelengths from 3 to 5 μm) and could alternate observations and calibration sequences. Alan Pickup, who had returned from Hawaii in 1987, was in charge of the CGS4 software. He says that the software was *'Done under the ADAM system and that was when I invented execs and configs. They were originally for testing so we could run a series of unattended tests and log the results overnight. It was conceived for repetitive testing and then I realised that these would be useful for real observing'*. Once CGS4 was in operation this system would allow data to be taken very quickly and, even though the exposures were often quite short, CGS4 worked at perhaps 80 % efficiency a lot of the time.

The CGS4 software system did not however stop once the data was taken. Its CGS4DR data processing system, based on the well proven Figaro library, "understood" concepts like execution sequences and target identity (calibration, dark, arc, flat, object) and could present, in virtually real time, 1D- and 2D-spectra which could optionally include background-subtraction, wavelength calibration, atmospheric correction etc. Talking about it many years later Phil Puxley, who had been one of Matt Mountain's PhD students, described this as '*Absolutely ground breaking at the time*'. According to Matt '*The concept was actually developed on a plane ride back from Chile where Phil and I had been observing with a CTIO IR spectrograph. Infrared spectroscopy data can be very complex because of the OH atmospheric emission, the highly structured atmospheric transmission combined with all the flat-fielding and calibration issues. On that run in Chile we had to spend all night observing, and then the whole following day reducing the data to see if we had even detected something, so we could work out what to do the following night – we got very little sleep, and were never sure if we had really good data. On the flight back we decided we had to develop a different approach for CGS4 and came up with the whole concept for CGS4DR*'. The overall software package was managed by Alan Pickup and CGS4DR was written by John Lightfoot and Steven Beard. Once the instrument arrived in Hawaii, responsibility for CGS4DR passed to Phil Daly, a software engineer who had joined ROE in October 1990, arriving via Cardiff and Leicester before being posted to Hilo. According to him CGS4DR had a number of known bugs and a list of desirable features and he fixed most of them. Later he ported the code to the Unix operating system and developed a CGS4DR demonstration that was shown at various scientific meetings.

With so much at stake, and such a complex system to commission, CGS4 was one of the first ROE instruments that had a lot of pre-delivery testing and a proper calibration plan. For the first time ROE had to build a full size flexure rig to move this instrument around and ensure it didn't flex internally before it could be delivered to the telescope. This testing drew heavily on the experience of front-line UKIRT staff who could relate to how it would actually be used on the telescope. First light in the laboratory was on 24th April 1990 when the instrument was switched on for the first time with an array present. An argon arc spectrum was detected within a few minutes.

The instrument, and its associated electronics, was shipped to Hawaii in December 1990. Steven Beard recalls '*one of the items shipped out to Hawaii inside the CGS4 crates was a motorised Dalek! I seem to remember that the CGS4 shipping was very generously specified, and there was sufficient spare capacity for us to ship out warm clothing and recreational materials. I shipped out a thick mountain jacket and a badminton racket and someone else must have shipped out a Dalek.*'

After laboratory testing in Hilo the instrument was transported to the summit and installed on UKIRT. There, unlike the other instruments which mounted via the instrument support unit, the CGS4 cryostat was bolted directly to the UKIRT mirror cell to provide the required stiffness and structural integrity. First light on the sky was achieved on Monday, the 4th of February 1991 soon after twilight and, thanks to the efforts of the UKIRT engineering staff, was some 6 days ahead of the commissioning schedule. Steven Beard, who was there, wrote a short account for the April 1991 ROE bulletin and described 15 people standing or sitting, shoulder to shoulder, in a semi-darkened room and staring at glowing display screens. He wrote that the first detection '*Happened so quickly it caught everyone unawares* '. Joel Aycock set up the telescope on a bright star (BS1552) and a stripe appeared on one of the displays. According to Steven it was several seconds before Dave Robertson pointed at the screen and said '*That's a star'*. The target had come straight down the slit, first time. A total of 186 observations were made during this first night and both low and high resolution (echelle) modes were tested. Later that week, with external project scientist Bob Carswell in attendance, CGS4 took a spectra of its first 16th magnitude quasar. Bob Carswell later published this in the Astrophysical Journal with the entire CGS4 team listed as co-authors.

**Fig. 9.4** PhD student Suzie Ramsay watches CGS4DR perform its "magic" during the commissioning run (Photo Steven Beard)

The first observing period in 1991 was done on the basis of "shared risks" observing in which the instrument was made available to visiting astronomers before it was fully characterised and tested on the understanding that from time to time observing would be lost due to the need for further engineering work and testing. Indeed the first scheduled observer had five nights allocated and lost all of them due to instrument problems, but shared risks observing did ensure that astronomers would get early access to the instrument and that its various observing modes would be tried out under realistic conditions very early in the commissioning process. It was, however, not without its difficulties, one such user recalls '*Observers had to use the engineering software to run the instrument and see the spectra produced in near-real-time. It was quite unfriendly, and I could never remember the arcane commands needed. Indeed, the support astronomer would only memorize the few settings that would work to get spectra printed out, and so it was difficult to know what results one was actually achieving at the telescope, and if something happened after the support astronomer had gone down the mountain the observing team was flying blind in terms of checking the spectra one was getting'*.

Matt Mountain and Suzie Ramsay stayed in Hawaii during the first few months of observing before they both returned to Edinburgh in the summer of 1991. They left CGS4 in the care of Gillian Wright who would be the CGS4 instrument scientist for the next six years during which various new operational modes, including spectropolarimetry, were brought into use. However, it was something of a baptism of fire, for before the commissioning was finished the detector in CGS4 suddenly stopped working.

It was a difficult experience, not only was the success of the instrument important to UKIRT and ROE, but there were perhaps inevitably suspicions from some staff at ROE that their colleagues at UKIRT had somehow broken their beautiful instrument. In fact the detector had been destroyed by an electrical spike, but it turned out that this was due to a grounding problem rather than an act of carelessness. It was really an issue that had not been understood about the vulnerability of that family of detectors and their controllers. It might have been regarded as a design failure, but it might equally be said that it was a feature of how sensitive those very new detectors were to voltages in certain places and that this had simply not been appreciated. However it was not a happy experience and one of the people involved puts it this way '*There was not much team working between the JAC people and some of the ROE people in trying to solve the problem*'. So there was a rather tense period of a few weeks while the CGS4 team tried to understand what had happened. Then, once the cause was fully understood, it was considered safe to put a new detector. There was only one device available, and that had a crack in it which spoiled its cosmetic properties, but aside from the crack it was a better detector than the one in IRCAM. In particular it had a very stable dark current which was good characteristic for use in a spectrograph. So the cracked detector went in and CGS4 went back into operation. Despite this difficulty Gillian thinks that '*On the whole the CGS4 commissioning went very well- in many ways it was the first example of the ROE building an instrument and then delivering it to somebody else'* (In the sense that by then UKIRT operations had made the transition into being more standalone and organised in its own right as part of the JAC, rather than as being an ROE group).

CGS4 was an immediate success and soon started to produce a large number of publications. These increased further when its cracked $58 \times 62$ element array was replaced by a 256 square detector in 1995. This detector upgrade, plus some mechanical improvements, made target acquisition simpler and provided better spatial and spectral resolution and improved sensitivity. Re-commissioning of the instrument with its new array, read out by a version of the improved IRCAM controller ALICE took place at the telescope on the 22nd and 23rd of April and between the 1st and 3rd of May. Despite some poor weather in the first part of the run the various modes of the instrument were tested and the upgraded CGS4 was put back into use.

Tom Kerr arrived as a new support scientist in September 1996. He had taken a PhD at Preston under the supervision of Andy Adamson on the subject of interstellar ices and had made his first trip to UKIRT in May of 1991, using CGS4 to study the rho Ophiuchus dark cloud. After finishing his PhD in early 1994 he moved to Nottingham where he worked with Peter Sarre on studies of the diffuse interstellar bands, a topic which drew him into optical astronomy, travelling to the Anglo Australian Observatory in Siding Spring as well as to Mt Stromlo, Kitt Peak and the South African Astronomical Observatory. Tom suggested to Peter that they do some infrared work, and so he found himself observing at UKIRT over Christmas in 1995. During this visit that Tom Geballe suggested that he would make a good support scientist and urged him to consider applying for one of the vacancies soon to be advertised.

**Fig. 9.5** Matt Mountain and CGS4 on UKIRT (Photo Steven Beard)

With Gillian Wright already planning to return to Scotland, Tom was earmarked to replace her as CGS4 support scientist and this came at an interesting time. There had been internal mechanical problems with the wheels that held the instrument's slits and CGS4 was soon in pieces as the JAC team worked on the problem. This gave Tom an early baptism of fire on instrument work and enabled him to gain at least some practical knowledge of how the instrument worked. Nonetheless since he had almost no experience of instrumentation work in the past he admits to being *'nervous'* at the thought of being responsible for UKIRT's largest and most complicated instrument. For all his nerves Tom would remain in charge of CGS4 until 2001 when he became project scientist for Michelle. The CGS4 role was taken by Paul Hirst in early 2001, but Tom was to remain a key player in the CGS4 team and indeed at UKIRT itself, for many years.

CGS4 dominated demand for UKIRT instruments for almost a decade, producing data for almost 500 publications, and often getting 60 % of the time on the telescope each semester. The instrument might have been used for an even greater fraction of the time but for the need to open it up from time to time to switch grating and internal optics and by the limited number of staff available to support it.

# Chapter 10
# Changes at the Top

## The Royal Observatories Merge

The arrival and commissioning of CGS4 was accompanied by a series of other changes as UKIRT began to settle into its second decade of routine operations. Some of these directly related to UKIRT itself, others were more remote, but many would have long-term implications for the way the telescope was run. The first big change came at the end of 1990 when Malcolm Longair, who had been director of the ROE for ten years, left to take up a position at the University of Cambridge. His departure marked a period of uncertainly for the ROE, which began when he was replaced by Paul Murdin, as Acting Director. Murdin's long association with the rival Royal Greenwich Observatory made him an unpopular choice with some of the ROE staff who feared a creeping RGO take-over of ROE or even the closure of the Edinburgh site entirely. The situation was only partly resolved when the two observatories were harmonised into a single structure in 1992. RGO director Alex Boksenberg, was appointed as head of the combined Royal Observatories and took up his new post on the 15th of March 1993. At about the same time Ian Robson, by now a Professor and the head of the Physics and Astronomy Department at the University of Central Lancashire, replaced Richard Wade in Hawaii as Director of the JCMT.

## Total Eclipse

Amongst all the staff movements, the summer of 1991 saw a rare astronomical event, a total solar eclipse visible from a major ground based observatory when the Moon's shadow passed across Mauna Kea. To celebrate this fortuitous co-incidence Kevin Krisciunas gave a number of public talks and, on the 28 and 29th of June, the University of Hawaii in Hilo was the site of two performances of a musical called



J.K. Davies, *The Life Story of an Infrared Telescope*, Springer Praxis Books,
DOI 10.1007/978-3-319-23579-0_10

"Total Eclipse" written by Kevin and his friend Margaret Harshbarger. This was not Kevin's first sojourn into the theatrical aspects of astronomical outreach, during the 1985–1986 apparition of Halley's comet he had dressed himself as Halley (complete with green tights) and taken a comet making demonstration around local venues.

The eclipse itself was on the 11th of July and Kevin and Dolores were at the UKIRT dome waiting for the night to end. The eclipse was not long after sunrise and, because UKIRT could not be aimed lower than 30° elevation, it was the only large telescope at the summit not being used for solar experiments. So, unburdened of the need to do professional astronomy, Kevin took a roll of colour slides of the eclipse with his own 6 in. (15 cm) telescope. Steven Beard had visited UKIRT just before the eclipse and remembers that a BBC Horizon programme was filmed on Mauna Kea and featured several UKIRT staff including a scene of Kevin '*Driving down from the summit singing songs from his production*' although in fact the song Kevin was singing was from one of his earlier works, the "Hilo All Star Revue" in 1987.

## A New Director

In late 1993, after nine years in charge of UKIRT and later as head of the JAC, Malcolm Smith left Hawaii to become director of the Cerro Tololo Inter-American Observatory in Chile. He had had a successful tenure and in the UKIRT newsletter Tom Geballe remarked that under Malcolm's leadership UKIRT had matured into the world's best and most productive infrared telescope. Tom also noted Malcolm's ability for low-key leadership, which had both maintained the UKIRT staff as a well motivated team in Hawaii while at the same time he was arguing UKIRT's case for funding with the scientific community in the UK.

With Malcolm's departure Ian Robson became Director of the JAC as well as of the JCMT and appointed Tom Geballe as Head of UKIRT. Tom wasn't sure about this promotion as he felt he was fundamentally an observer and management wasn't something that he felt he had a lot of experience of, or was that keen to take on. However, in reality he had been gradually developing those very skills. Tom had remained at UKIRT after the Dutch withdrew in 1987 to concentrate on their involvement with the JCMT. With the onset of JCMT operations Malcolm Smith had delegated the day-to-day running of UKIRT, and the supervision of the UKIRT support scientists, to him as "Astronomer-In-Charge". At the time Tom said that he found himself in the *'Strange position of being an administrator'* and that he preferred to think of himself as *'An independent scientist rather than some-one in charge of persons, places or things'*. Recognising that he would henceforth be representing UKIRT at meetings around the world he went on to ask for the guidance of his colleagues in order that he could do this *'Properly'*. He must have succeeded in this because as the JAC evolved, his responsibilities increased still further and he became an Associate Director for UKIRT.

Thus Tom had been effectively running UKIRT on a day-to-day basis for some time and so Ian persuaded him that the change would be good for UKIRT and probably for Tom as well. Nonetheless at heart Tom remained an observer and not a manager. He admits that he allowed himself to be promoted partly out of a sense of duty and he worried that he would no longer be *"One of the guys"*. He wanted to stay in his old office along the corridor with the rest of the support scientists, but was persuaded, or instructed, to move into the management suite, isolated by the physiological barrier of secretaries Anna Lucas and Donna Delorm. He recalls coming out of his manager's office, seeing a group of UKIRT scientists talking together down the hallway, and thinking *'Yes - I should go down and join them, shoot the breeze, and find out what's up'* and then thinking *'No I shouldn't, they might be talking about management or, even worse, about me, and it could be awkward if I approached them'*. To avoid the conflict he walked downstairs and along the ground floor hall, rather than go past them. In fact Tom need not have worried; he continued to work on the mountain as a support astronomer and remained "One of the guys". Tom was a very popular leader, perhaps because he had no pretentions about being a manager, and as a result most of the staff worked hard to make sure things ran smoothly and protected him from any hassle.

While all of this was happening there were other changes. Two long serving members of staff departed in the spring of 1992. First to leave was David Beattie, who retired in February after more than 25 years with SRC and its successor organisations. He had been at UKIRT since 1979 and had established a long record of commitment to the telescope and its staff which had been a major factor in its success early on. Next to go was Yolanda Boyce who departed in March after many years at the JAC, first with UKIRT and later with the JCMT. March also saw the departure of Mark Casali for a position at the ROE in Edinburgh. The intention was that Mark, who had been the UKIRT scientist in charge of IRCAM for several years, would replace John Davies as the ROE nominee to join the ground station team for the ISO satellite while John came to Hawaii as a support scientist. After a period over of overlap with Mark in Edinburgh, John arrived with his family in May 1993. There were other staff movements too. In January 1992 Matt Mountain arrived to work on plans for improving UKIRT with active or adaptive optics, but his tenure was to be short lived, he left again in November for a position as Gemini project scientist. Phil Williams, who had been in Hawaii since New Year's Eve 1983 and had risen to Acting Chief Engineer departed, as did Bill Duncan who returned to Edinburgh to work on JCMT instrumentation.

## Unexpected Celebrity

Phil Daly, who would remain in the software group until 1997, had an unexpected encounter with TV celebrity. He, along with Tom Geballe, Gillian Wright and Alan Bridger were at UKIRT one day when the BBC turned up to do some filming for one of their science programmes. After Tom had been filmed talking in the dome the group were asked to do something for the camera. Phil brought up some sort of image on the screen and then he and Gillian began to act their way through a little scene. The director stopped them about half way through to ask if they could do the same thing but with a different desktop-background image because Phil used "Mickey Mouse in Fantasia" and the producer was concerned that if someone from the Disney Corporation saw the screen then they might sue. So, they changed the image, did the little acting thing again and everyone was happy. The BBC left and the UKIRT team got on with their night's work.

About 18 months later Phil was spending a night in the UK en-route to a meeting. In the evening he went to the local shop for some beer and while he was looking at the shelves a man entered with his young son. The child asked "Didn't I see you on telly last Saturday night?" Since Phil had just flown in from Hawaii that day this seemed so ridiculous that he told the child to go away. However, the next morning in the laboratory someone asked, "Hey, Phil, didn't I see you on telly last Saturday night?" Well, that was too much of a coincidence and, as it happened, the BBC programme had just been shown and the child in the shop was right.

# Chapter 11
# Upgrading the Telescope for the 21st Century

## A Man with a Plan

UKIRT had been designed as an inexpensive telescope, but in some respects these cost saving decisions turned out to be advantageous in the long run. For example, its thin and hence flexible, primary mirror was, thanks to the commitment of that £9325 of extra spending in 1977, of a quality comparable to an optical telescope. Furthermore, to save money, the original planners had opted for the smallest possible dome. Even though the dome design was later increased in size to allow flexibility when mounting alternate telescope top-ends, it was still a very tight fit. The dome was never intended to be actively ventilated, but would eventually turn out to be comparatively easy to modify. Finally, the telescope's deliberately lightweight structure had lower thermal inertia than any similar-sized telescope and this, if properly exploited, could aid temperature stabilisation which was a known factor for improving image quality. However, even after a decade of operations UKIRT had many problems. The pointing was not very good, and once on target, its tracking was unreliable. The small dome offered little protection from the wind and, until the advent of the new windblind went someway to improving the situation, the telescope would shake if pointed into even a modest breeze, severely degrading the image quality being produced by the excellent mirror. Alignment of the optical elements was poor and a turned down edge in the secondary mirror, which allowed the main beam to "see" the heat emission from warm structure surrounding the secondary, led to poor performance in the thermal infrared. Finally, warm air plumes rising from the ever increasing number of electronics racks underneath the mirror further degraded the local image quality.

The result was a telescope that rarely delivered the high angular resolution images that its optics, and the conditions on Mauna Kea, were potentially capable of delivering. This was hardly surprising since the telescope had been designed when image quality was almost irrelevant for single element photometers and when even 58 × 62 pixel infrared cameras were a decade away. However, towards the end



J.K. Davies, *The Life Story of an Infrared Telescope*, Springer Praxis Books,
DOI 10.1007/978-3-319-23579-0_11

of the 1980s the scene was changing, infrared cameras had arrived and were getting better all the time. Furthermore, active and adaptive optics were promising much better imaging from ground-based telescopes in the near future. If UKIRT could not adapt to this changing world, its future might be rather limited. Then, in September and October of 1988, there was a six-week shutdown for dome repairs and enhancements which left the support team, amongst them Tim Hawarden, with some free time. As Tim put it *'The Devil makes work for idle hands, and I spent much of that time thinking about the possibility of solving as many of UKIRT's problems as possible'*.

The first task was to divide up the potential areas for improvement and to try and understand what the situation actually was in each of these areas. Only once this was known could plans be developed to improve on them. Determining the optical parameters of the system turned out to harder than expected, so much so that a self-consistent description of the telescope could not easily be put together. This was a major impediment to further progress and had to be tackled before it would be possible to decide what optical improvements could be made. Evaluating the thermal properties of the dome was far easier and was done by the simple expedient of attaching mercury thermometers to the structure with sticky tape and recording the data by hand. Although crude, these measurements indicated significant areas for potential improvement and also allowed a justification for spending some of UKIRT's limited budget on a thermal sensor and data recording system more sophisticated than a couple of thermometers and a man who walked around from time to time with a notebook. Telescope pointing and tracking, always a weakness of the early UKIRT, was improved by tweaking the pointing model.

In September 1989 the top-end of the telescope was removed and some modifications were made to improve the alignment of the primary and secondary mirrors. The first attempt failed due to what Tim described as *'A 14,000 ft type blunder'* but the second attempt succeeded and soon the telescope was producing images much more in line with those expected from an optical telescope than an infrared light bucket. Most importantly, some of the remaining optical problems were understood and could be removed by taking advantage of the once innovative 72 pad pneumatic support system and using it to actively control the shape of the mirror.

In mid-1991 some infrared pictures were taken, which, after allowance for the relatively large scale of the pixels in IRCAM, suggested that the images had a Full Width Half Maximum (a standard measure of telescope image quality) of about 0.41 arcsecond. This was almost as good as achieved by any ground-based telescope and was an extraordinary endorsement of the polishing done by Grubb Parsons some 15 years earlier; more than justifying the risks taken in committing the extra money for which they had asked.

## The Upgrades Project

In August 1991, Tim Hawarden, buoyed by the success of the first stage of telescope improvements, presented a detailed plan for an upgrades project and circulated it to the UKIRT engineering team. His plan included a suite of possible activities in the short-, medium- and long-term. The first stage was to correct problems with the secondary mirror support system, which tended to flop about when the telescope was pointed far away from the vertical. This was to be followed by a comprehensive campaign to investigate the optical and mechanical behaviour of the whole telescope. Only once this was done would it be possible to decide how the remaining problems could best be tackled.

In the medium-term, efforts would be made to improve the stiffness of the telescope by enhancing the control system's servo motors. This had been a longstanding problem, Andy Longmore and Bill Parker had worked on the telescope drive servo for years, but they could not get good performance over the whole sky. If the servo was adjusted to be stiff to windshake in one part of the sky it tended to oscillate in others. Fixing this issue would reduce the susceptibility of UKIRT to windshake, with a corresponding improvement in operational efficiency. Additional information gathering would be done by the installation of a seeing monitor, which would provide the data required for rational decisions about potential optical, mechanical and dome related improvements.

In the much longer-term, further enhancements would require the installation of some sort of low-order adaptive optics system, most probably in the form of a new multi-axis controllable secondary mirror and telescope top-end. This new secondary mirror could be polished to cancel out the residual optical errors in the primary and be supported in such a way as to significantly reduce the thermal emission from the top of the telescope, improving performance in the thermal infrared. Having improved the secondary mirror and its pointing system it would then make sense to improve the support system of the primary mirror to remove various aberrations (notably astigmatism), which resulted from the inadequacies of the original mirror cell. Finally, to take advantage of these optical changes it would be necessary to institute a programme to improve the quality of the atmosphere in the dome itself. In particular the heat generated near the mirror by the scientific instruments attached to the mirror cell needed to be removed and the temperature gradients which built up in the structure during the day, and which produced "dome seeing" once the shutters were open and the telescope exposed to the cold night air, had to be minimised. Active control of the dome temperature was also considered, but simple calculations showed that the costs of the electricity involved, about $200,000 per year, would be prohibitive.

These ideas were set out in some detail in a paper presented to the UKIRT Users Committee on the 23rd of September 1991. This, as well as setting out the technical arguments reported above, included some cost estimates and calculations of how the improvements would generate improved telescope effectiveness (i.e. better images require shorter integration times to reach some given signal to noise ratio). The basic suite of simple improvements would cost about $570,000 to carry out and would improve the telescope effectiveness by about a factor of four. Adding the active optics, i.e. the primary mirror control and a fast tip-tilt secondary mirror, would cost a further $800,000 but would result in an effectiveness gain of about a factor of 10. This suggested a programme with the following goals.

1. UKIRT **must** not degrade the delivered image quality by more than 0.25 arcsecond (measured as FWHM). This was the then predicted image size for a tip-tilt corrected 4 m telescope at K (2.2 μm) in median observing conditions on Mauna Kea.
2. UKIRT **should** not degrade the delivered image quality by more than 0.12 arcsecond (measured as FWHM) (this is the diffraction limit of UKIRT in the K band)
3. Local seeing effects should not appreciably degrade image quality; although no specific target for dome seeing was set.

The project was approved by the UKIRT Users Committee in September 1991 but there was a problem, the SERC did not have sufficient money. Most of the improvements could not come from the routine running costs of UKIRT, significant extra funds would be required and these would have to be requested in competition with other projects from across the SERC portfolio. It was not a good time for public spending and government cutbacks were in the wind. The money to upgrade the telescope was not likely to be available unless it could be found from outside the SERC.

Luckily, German astronomers, notably from the Max Planck Institute for Extra-terrestrial Physics (MPE) in Garching and the Max Planck Institute for Astronomy (MPIA) at Heidelberg were interested in gaining access to UKIRT. The German groups were involved in instruments for the ISO infrared satellite and they saw UKIRT as a telescope that could provide useful ground-based support to their ISO observations. Exploratory talks between the UKIRT unit of the ROE and the Max Planck Society had been taking place on and off since early 1987 to establish how a deal might be struck that would enable both parties to get what they wanted while protecting their own interests. The option of simply selling time to the Max Planck society was mooted, but there were a number of issues with this. Not the least was that under the UK's policy for allocating telescope time to the best proposals irrespective of their national origin, the Max Planck groups might have got some time for free under the existing rules and if they paid, and other nations did not, this would create obvious political difficulties. There was also a question of how a night of observing time might be priced, for example should it take account of the initial investments or just the annual running costs? What was being sold, a night on bare telescope or access to the entire UKIRT instrument suite and the support staff required to operate them? What about instrumentation developments? As it happened the Germans were interested in high resolution infrared imaging and their desire to do this matched neatly with the UKIRT team's plans to upgrade the telescope if only they could find the money. The win-win scenario was clearly possible and soon the talks had moved on from a supplier-customer relationship to a partnership in which the Max Planck society would contribute technical effort and hardware in return for observing time.

Over the first half of 1992 an agreement was hammered out between Paul Murdin, then the acting director of the ROE, and his counterparts at the MPIA and MPE, Steve Beckwith and Reinhard Genzel. Under this agreement MPIA and MPE would provide various hardware elements for the upgrades project, notably a new top-end with a fast, tip-tilt secondary mirror, in return for guaranteed observing time. Possible contributions towards a far infrared camera were also postulated but in the end did not feature in the final arrangement. After rejecting the idea that the German share could be based on some evaluation of the performance gain produced by the new equipment, it was agreed that the amount of time to be allocated to the Max Planck astronomers would be based on the relevant financial contributions of SERC and MPIA/MPE to the cost of UKIRT operations during the period of, and for a time after, the upgrades project.

Of course, this ratio was not a simple thing to calculate, different organizations have different ways of calculating costs, which had to be somehow reconciled. Furthermore, hardware development is often more expensive than expected and the UK side was concerned that runaway costs in Germany might result in them gaining considerable observing time simply because their side of the hardware suffered cost over-runs which would automatically translate into more observing time. There was also the issue of who would own the mirror and other equipment provided by Germany when the agreement ended, would it become the property of the UK or remain that of the Max Planck groups? Without UKIRT the Max Planck Society had no use for its tip-tilt secondary, but without a secondary UKIRT would be useless! It was not a problem to be taken lightly in the minds of the administrators who would ultimately have to approve the deal. At least once the agreement between Murdin and Beckwith seemed near to completion when SERC's central office objected to some of its provisions and set back the process. By December 1992 these difficulties were mostly ironed out and a deal was struck that would make the Germans partners in UKIRT for a time.

The MPIA in Heidelberg officially joined the programme on the 21st of September 1993 by signing an agreement that became effective from the 1st of October. The MPIA agreed to provide a new top-end, which employed a light-weighted mirror by Prazisions Optik (Germany), controlled by a system provided by Physik Instrumente (Germany). Control was in two stages: a 6-axis "slow" hexapod positioning system and a 3-axis "fast" piezoelectric tip-tilt system with a driven counterweight to minimise vibration transfer to the telescope structure.

The agreement was for a multi-year collaboration eventually involving four groups: ROE, RGO, MPIA and the JAC. The overall project manager (from 1993) was Donald Pettie at the ROE and for most of the programme the core group in Hawaii comprised Nick Rees (optics, computing, control systems); Tim Chuter (electronics, optics, operational issues) and Chas Cavedoni (JAC Project Manager, Mechanical Engineering) who was "poached" from the nearby IRTF telescope to work on UKIRT. Although Matt Mountain was the original programme scientist, with his departure to Gemini in late 1992 Tim Hawarden took on the mantle of local scientist. Considerable support was provided by other JAC staff and from the ROE in Edinburgh with about 30 people involved at some or other stage of the effort. For example in 1992 a team from the ROE's applied optics group, led by the same Colin Humphries who had gambled on the extra polishing 15 years earlier, came to UKIRT to make a series of interferometric tests on the mirrors. These went well and allowed the optical engineers to separate out the effects from the primary and the secondary mirrors discovering that the main aberration in the primary was astigmatism, which had the potential to be fairly easily correctable.

UKIRT was not the only telescope on Mauna Kea that was being modernised and Tim is the first to admit that the upgrades team '*Freely plagiarised from the IRTF upgrades programme. For several years we let them make the mistakes first, but then, perhaps mistakenly, we overtook them and had to start making our own blunders'*.

## Almost a New Telescope

In 1995 the support system for the primary mirror was completely upgraded to be able to maintain the mirror figure to better than 100 nm in all orientations of the telescope. The new system comprised a three-sector axial support with about 80 pneumatic cells. Twelve actuators were installed around the mirror mounting cell to push and pull the mirror into whatever shape was desired in order to correct the aberrations that had been degrading the image quality for so long. Initially these adjustments were made manually until sufficient tests over the whole sky had been accumulated for a computerised automatic primary mirror control system to be used.

The new secondary, mounted on a new top-end with a much stiffer structure was installed in the first half of August 1996. At the same time the instrument support unit at the bottom of the telescope was fitted with a new CCD based fast guiding system mounted on a new crosshead. The fast guider employed a sophisticated Kalman filter, and was capable of guiding on an image with only about ten counts/cycle. The guider also provided a wider-field (30 arcsecond) acquisition mode, and a Shack-Hartmann option for autofocus and measurement of seeing.

Dome seeing, which arises from air currents resulting from temperature gradients within the building, was controlled in a number of ways. The telescope volume was isolated from external heat sources by insulating the common room ceiling underneath and a vestibule was built around the cargo entry door at dome level. Ventilation was improved by inserting 16 louvred and shuttered ventilation ports around the lower edge of the dome. A prototype was installed in February and March of 1996 with the remainder being added that autumn. When opened these vents allowed air flow through the dome, and permitted actively flushing of the dome space using the coudé room below as a plenum chamber, with extractor fans in the plant room drawing out the warm air. When these fans were turned on, air flowed down through pre-existing holes in the dome floor, downstairs to the coudé room and then to the outside. These ventilation ports also dramatically changed the look of the dome from the outside, making it vaguely resemble a flying saucer. Mirror cooling was provided by a large, low vibration fan of a type used in submarines which was designed to run very quietly so that, as Tim put it, *'Other submarines cannot hear them'*.

An intensive period of testing and shaking down the new system then began. While overall the new light-weighted secondary performed well it was soon apparent that the mirror showed "spectacular" print-through effects resulting from the removal of material from its back face to lighten it. This quickly became known as the "waffle pattern" and although serious in the longer-term it did not produce a major effect on the actual images. The secondary mirror's outer edge was already known to be slightly turned-down and there remained some other residual optical problems which were at the limit of the primary mirror system's ability to solve. There was, however, a big improvement in the pointing; an improvement of almost a factor of 10 was seen in the first test. This probably resulted from the tightening of bolts at the top of one of the telescope trusses, which had been found to be only finger tight. The first use of the tip-tilt system was on the night of the 14th of August. Within a few more days various adjustments of the primary mirror control system by Nick Rees had reduced the optical aberrations (astigmatism and trefoil) and produced images that seemed to be close to the theoretical best that could be expected.

**Fig. 11.1** Installing the dome ventilation apertures was a tricky business in the cold, thin air of the summit (Photo Chas Cavedoni)

**Fig. 11.2** UKIRT with its ventilation louvres which drastically changed the shape of the dome (ROE Photo)

One thing that was planned but not implemented was insulating the entire dome floor to limit heat coming up from the rooms below. The project was about to place four inches (10 cm) of foam-backed board over the whole dome floor when reports of a glycol fire at the Cerro Tololo Inter-American Observatory in Chile caused them to rethink this idea. The flammability of various insulating materials at altitude was tested by the nearby Keck observatory, and was found to be higher than at sea level for almost all materials! Any glycol spillage from the proposed mirror primary cooling which soaked into the foam would thus lead to an unacceptably high fire risk and so this plan was abandoned. It was replaced by a plan to insulate the ceilings of the warm rooms underneath the dome floor.

The mirror cooling itself was installed and first tested in 2000, but, for a variety of reasons, was not commissioned for a number of years. One of the causes for the delays was that some of the plumbing for the cooling was prone to leak and ran underneath the primary mirror where it was not easily accessible, another was simply the load on the engineering team which had many other tasks, including preparing the JCMT for the arrival of the SCUBA-2 instrument, to perform. By mid-2007 Andy Adamson would write that it was hoped that the system would be operational by the end of the year but in fact it was well into 2008 before it became operational. Thor Wold recalls that he started a bit of an underground campaign to get the system implemented by the expedient of mentioning, apparently casually, to politically connected visiting observers words to the effect that *'It sure would be nice if we had the cooling system running, I'm sure that if we did the seeing would be better than this…'* When the system was finally brought into use it was what Tom Kerr called *'A great success, it allowed us to observe projects that require good seeing early in the night which was not possible before'*. Thor agrees with this assessment saying that the difference when it was turned on was *'Astonishing'* and that it *'sure does make a difference at the start of the night'*.

Overall the upgrades programme had been a huge success. The pointing and tracking was greatly improved and the tip-tilt fast guider greatly improved UKIRT's image quality by solving two longstanding problems, the telescope shaking in the wind and the wobbling as the telescope moved in Right Ascension. These motions still occurred of course, but the fast guiding system was much faster than the shakes and wobbles and could correct for them almost the instant they happened. The result was better images all over the sky and in almost all conditions. However, the problems of print through and the turned-down edge of the new mirror could not be easily solved and would eventually require the manufacture of a new secondary mirror. This would be expensive and so the agreement with the MPIA was extended, granting them more guaranteed time in return for the cost of a new secondary. The new mirror, which was made slightly oversized and with the edge then ground off to avoid the turned down edge problem, was produced to a very high standard. It was installed on UKIRT on 14 June 1999. It showed no signs of the problems of its predecessor and by the end of the project UKIRT was producing images which, in Tim's words *'Brought UKIRT to a pinnacle of performance which had been unimaginable to the original designers of the telescope'*. The upgrades programme had been a stunning success at both technical and personal levels. Steve Beckwith recalls that *'There had been some creative and high quality German engineering as part of an international team spirit that was instrumental to the success of the project'*.

Tim Hawarden left Hawaii in the spring of 2001 to return to Edinburgh. The UKIRT newsletter issued soon after contained several tributes to his work on upgrading UKIRT. Tom Geballe wrote *'The upgrades team in Hawaii (including Tim Chuter, Nick Rees and Chas Cavedoni) that Tim led did an absolutely phenomenal job and should be given most of the credit for finding a sensible and workable route to allow UKIRT to jump [into] the new era of high image quality astronomy without missing a beat, and thus to continue as one of the world's most productive telescopes'*. Nick Rees himself added *'The simple truth is that the upgrades project was Tim Hawarden's idea, it was his baby and we were bonded together with some special sort of Tim glue. I don't know how he makes it, but it is extremely rare and springs from his eternal willingness to have a heated discussion on just about anything, tempered with his incredible support for his fellow workers and a true generosity of spirit that is unmatched in my experience'*.

# Chapter 12
# Restructuring

## Panels and Decisions

In late 1994, while the upgrades programme was in full swing, there was a review of the UK programme in Optical, Infrared and Millimetre astronomy by a panel chaired by Professor Jim Hough from Hatfield This reported in January 1995 and made a number of far-reaching recommendations which included a further rationalisation of the two Royal Observatories into a single, smaller, organisation to be called the UK Astronomy Technology Centre (UKATC). It also recommended the devolution of the island sites in Hawaii and La Palma from their home bases at the, soon to vanish, Royal Observatories. For staffing purposes these now independent overseas sites would no longer be a part of the Particle Physics and Astronomy Research Council (PPARC), into which the astronomy area of SERC had now moved after a reorganisation of the UK research councils, and their staff would be locally employed. A consequence of this arrangement would be that the rotation of personnel between the UK and the telescope operations teams on the islands would stop. This created considerable uncertainties at both the island and UK sites and did nothing to speed progress on new instruments. On the positive side, the Hough report did describe UKIRT and the William Herschel Telescope in La Palma as the top priority UK telescopes until at least 2005 and identified potential instruments (which would become UFTI, UIST and Michelle) to provide a stable suite of permanently mounted UKIRT instruments.

The conclusions of the Hough panel were endorsed by PPARC at the highest level and it was clear that there would be major changes at all the sites affected. Inevitably the implementation of the panel's recommendations required yet more discussions, creating many more months of anxiety and uncertainly for staff in Hawaii and at the UKIRT Home End in Edinburgh. In what became known as the "observatory wars" the Royal Greenwich Observatory, by now located in a new building in Cambridge, and the ROE in Edinburgh slugged it out to see which would survive as the host of the new UKATC. UKIRT staff in Hawaii, especially


J.K. Davies, *The Life Story of an Infrared Telescope*, Springer Praxis Books,
DOI 10.1007/978-3-319-23579-0_12

those on rotation from Edinburgh who could do little to influence events, watched with some trepidation. To them ROE was the stronger and most obvious choice but when, they asked themselves, had the UK power brokers, dominated by Oxbridge graduates, ever closed something in the South of England in favour of something in Scotland? Jim Hough was directed to produce another report, this time on the optimum location for a UKATC and, in June 1995, his UKATC analysis panel came down supporting Edinburgh. It was however not the end of the story, the observatory wars would continue for several years.

While the battle between Cambridge and Edinburgh was fought out in the committee rooms of England, it was time to take forward the rest of the Hough report's recommendations. Yet another panel, chaired this time by Professor David Williams, was tasked with producing a costed plan for the development of the UK's existing optical, infrared, millimetre and radio ground-based facilities over the next 10 years. It was asked to '*Submit a report to Council through the PPARC Executive no later than 31 December 1995, to enable a development plan for the UK ground-based telescopes to be incorporated in the 1996 PPARC Business and Corporate Plans'*. Known officially as the Ground Based Telescope Development Panel it was also flippantly referred to by some as the "Great British Telescopes Destruction Panel". Ian Robson presented the panel with a well-argued plan for UKIRT up to 2005, which included instruments called UFTI, UIST + MOS, an upgrade of CGS4 and an infrared Multi Object Spectrograph. His document also mentioned options for IR wavefront sensors for adaptive optics, an adaptive secondary mirror and a wide-field imager. Ultimately, the Williams review was quite positive for UKIRT, in the short-term it effectively endorsed UFTI, Michelle and UIST, all three of which would be funded in PPARC plans and in the longer term it identified a role for UKIRT as wide-field infrared facility in the era of 8 m telescopes.

However, yet another review was about to start. In the mid-1990s the Conservative government had decided that as many functions of government as possible should be abolished or privatised, leaving government departments to do only what was absolutely necessary. They then established a process, which they called Prior Options, to ask "Is this function really necessary and if it is, can it be done by the private sector"? The Research Councils, and hence the Royal Observatories and their telescopes, were subject to Prior Options reviews and in July 1995 the Times Higher Education supplement ran the headline "Review frenzy sets Royal Observatories reeling". According to the report *'The astronomy community is reeling after the announcement of another review of the Royal Observatories, which will bring the number of reviews this year to three and continue a chain of studies that started in the early 1980s'*. Jim Hough was not amused, his panel had already evaluated, and rejected privatising the observatories. He was quoted by the paper as saying *"We are happy with the recommendations that we made and perhaps we are disappointed that our recommendations have to be delayed until the prior options is completed"*.

**Fig. 12.1** Despite the uncertainties over UKIRT's future, the staff continued to make Herculean efforts to maintain the telescope's reputation. Tim Carroll injured his back one night changing a tyre but the next night was back in the control room. He was only comfortable standing so he put his keyboard on a box and spent the night using a trolley as a backrest. It was typical of what made UKIRT special (Photo Richard Ellis)

Nonetheless, Prior Options was government policy and so the review had to be carried out. Hundreds of staff hours were spent writing reports and going to meetings, while the UKIRT staff in Hawaii were treated to the depressing sight of potential bidders for privatised contracts walking around making video tapes of their buildings. It was rather like being eyed up by an undertaker as a possible future client, and it did nothing to improve morale amongst people who were, by now, suffering from "review saturation". Ian Robson wrote in the March 1997 JCMT Newsletter *'Prior Options has, unfortunately, sapped effort at the JAC and some of the work that we had intended to do has slipped. It is to be hoped that the uncertainty surrounding this process will be soon removed and we can get back to concentrating on delivering high quality and cost effective science'.*

The uncertainly continued until, on the 4th of July 1997 (ironically independence day in Hawaii), the science minister of the UK's new Labour government announced that the Prior Options exercise for the observatories had been, in effect, abandoned and that PPARC could get ahead implementing the plans to proposed by the Hough and Williams reports 2 years earlier. So, on the 1st of April 1998, after almost another year of uncertainty and angst, the Royal Observatory Edinburgh ceased to exist as a legal entity, its staff being divided between the University of Edinburgh and the new UK Astronomy Technology Centre. This separation of the ROE and Hawaii meant that the UKIRT management of Ian Robson and Tom

Geballe found themselves reporting directly to PPARC at its Swindon headquarters and being seated on some influential decision making committees. It also meant that the budget would henceforth flow directly to Hawaii from Swindon which would require a higher level of financial management and support, a load carried partly by Ian Midson and local finance officer Lindsay Marcer. Fortunately the relationship with the new UKATC was excellent since the UKATC director Adrian Russell had been head of JCMT instrumentation under Ian in Hawaii some years before.

## Arrivals and Departures

One of the consequences of the break-up of the links between the UK and Hawaii was that there would henceforth be very few ROE staff coming to UKIRT on tours of duty. One by one, as the ROE staff either returned home or left completely if no jobs could be found for them at the UKATC, they were replaced by people employed under contract to the Research Corporation of the University of Hawaii. Amongst these was Sandy Leggett, who had been a research fellow at ROE in the mid-1980s and who arrived at JAC in August 1996 to take over responsibility for IRCAM following Colin Aspin's departure. Antonio Chrysostomou, from Jim Hough's group in Hatfield, came to support UKIRT's polarimetry projects and become editor of the re-launched UKIRT Newsletter when publication restarted at issue 1 in June 1997 after a gap of several years. There were new technical staff as well, Maren Purves joined the software group in early December 1996 and was joined by Cameron Mayor (electronics), who left in July the next year, and controls engineer Alvin Balius. Alvin did not stay long; he left in September 1997 to be replaced by Yaguang Yang who was in turn replaced by Russell Kackley.

Long-serving and experienced local staff were also leaving, often to go to the nearby Gemini telescope which was perceived to have better long term prospects following all the reviews and uncertainties about UKIRT's future beyond 2005. Kent Tsutsui, an electronics engineer at UKIRT for over 15 years, resigned in January 1997 to take a position with Gemini and Chas Cavedoni joined him after several years working on the UKIRT upgrades project. The 1st of April saw the departure of telescope operator Dolores Walther after 14 years at UKIRT. Gemini was starting to build up its operations team and made her what she calls *'An offer she could not refuse'*. It seems that Gemini planned to use an operations model that she wanted to see at UKIRT, where the telescope operator not only operated the telescope but was also trained for other duties at sea-level, allowing them to be more integrated into the science staff. Dolores says that her UKIRT position *'Didn't allow for anything but operating the telescope. I always felt that was a waste of talent'* so she left.

Dolores's departure left a large hole in the telescope operations team and led to the creation of a new experimental post that combined the functions of telescope operator, or Telescope Systems Specialist as they were by now called, and Post Doctoral Research Associate (PDRA). Two of the these PDRA/TSS post would be created, each would spend half as much time at the summit as the two remaining full time TSS (Thor Wold and Tim Carroll) and spend the remainder of the time doing their own research projects.

The first PDRA/TSS was Stuart Ryder, who had rescued a UKIRT article from his school wastepaper basket so many years ago. Stuart does not ascribe anything special to the UKIRT article, he says *'I wouldn't say that this article alone got me interested in astronomy, I still have scrapbooks with newspaper clippings on space from as far back as 1979 (at age 13) so it was just one of many items that took my interest. Until I saw the UKIRT job advertised in late 1996 I never gave much thought to observing on Mauna Kea, let alone working there. It could just be coincidence that one of my oldest space items is a story about UKIRT, but maybe it was a portent of my then far-future..'*. Stuart arrived in May 1997 and was joined in the other PDRA/TSS position by Chris Davis, a teacher who had made the jump into astronomy several years earlier. Chris started in July 1997.

## UFTI + TUFTI

After various preliminary studies in 1996 it was proposed to build a new near-infrared imager for UKIRT to take advantage of the larger format detectors that were then becoming available. The new instrument was to be based around the 1024×1024 HgCdTe Hawaii detector which was sensitive to wavelengths from 0.8 to 2.5 μm. As such the new instrument's operating range was smaller than that of the existing IRCAM, but with an image scale of 0.09 arcsecond/pixel and an overall field of view of 90 arcseconds, it would be able to take full advantage of the improved image quality delivered by UKIRT after the telescope upgrades programme. There was a strategic need to develop the instrument quickly even if it would have a short lifetime before it was replaced by a more capable facility in a few years. This meant that the design would have to be a simple and cheap camera, so it was known initially as BROWNIE a name soon changed to the UKIRT Fast Track Camera and subsequently amended to the UKIRT Fast Track Imager (UFTI).

The original design was largely the work of Peter Hastings and Eli Atad of the UKATC, but in the end the project was carried out by Pat Roche and Phil Lucas at Oxford University where the instrument was constructed. To achieve a speedy delivery it was planned to use existing designs wherever possible. The limited space between CGS4, the largest instrument normally mounted at the UKIRT Cassegrain focus and the forthcoming, and even larger, Michelle required a tall and narrow cryostat. Filter wheels contained the usual narrow band and JHK filters and two non-standard broad band filters (*i* and *z*) to exploit the short wavelength sensitivity of the detector. IRCAM3 would be retained for observations at wavelengths between 3 and 5 μm.

The original plan was to commission UFTI in May 1998, but the schedule gradually slipped and UFTI was eventually taken to the UKATC in June for acceptance testing. While the software integration was successful, a serious hardware issue with the science grade detector array was discovered. The problem was a faint glow in one quadrant of the array, which was at first attributed to a subtle light or heat leak, but which persisted despite all efforts to eradicate it. When it became clear that it was in fact a hardware problem, probably a defective wire, the manufacturer agreed to supply a new science array free of charge.

UFTI was shipped to Hawaii on the 2nd of September 1998 with the replacement science-grade array installed and with the last remaining hardware problem, condensation on the large cryostat window, solved by attaching a small fan next to the window to blow a stream of air across the glass. First light was on the 29th of September 1998 and although successful, the next few months were plagued by poor weather and filter motor problems. In January 1999 UFTI was brought down to Hilo and the motors repaired by the JAC staff. Soon, the camera began to routinely produce data which, in terms of area coverage and sensitivity, was considered to be extremely competitive with other infrared imagers on Mauna Kea.

Once it was determined that UFTI was fully functional and could be relied on as a workhorse 1–2.5 μm instrument, UKIRT staff modified one of the IRCAM cryostats to improve its performance in the thermal regime. The existing warm fore optics were replaced with a new cold snout to reduce the thermal background and to match the pixel size to that of UFTI. Although the field-of-view was reduced to 23 arcseconds, the resulting lower background levels enabled thermal data to be taken much more efficiently, and the smaller pixels meant that the images could be fully sampled. The modified IRCAM was referred to as TUFTI, or Thermal UFTI, a name coined by engineer Tim Chuter. Although Tim had a long history of involvement with IRCAM, and so might be considered to have naming rights, it was not a title that found favour with at least one of the support scientists who felt that naming a UKIRT instrument after a stuffed red squirrel from children's television lacked dignity. TUFTI was commissioned in the autumn of 1999.

# Chapter 13
# ORAC: It Pipes, Therefore It Is

## ORAC Is Conceived

The ideas which crystallised into the ORAC project developed gradually at the JAC during the mid-1990s. Key to the process was probably Alan Bridger, head of the UKIRT software group, who had been thinking for some time about how to radically improve UKIRT's end-to-end software system to make it easier to maintain and upgrade. In this he was not alone. He says '*The ideas/thoughts behind what would become ORAC had been evolving while I was still at the JAC, the semi-automation of observing, the evolution of CGS4/IRCAM_PREP into UKIRT_PREP, the CGS4_DR system and so on. Various things had been steadily coming together over a few years, and from a number of directions and individuals*'. There was also the issue of the control, user interface and data reduction for the new instruments which were on the way. For example, UFTI was being produced under the then popular 'faster, cheaper, better' philosophy and was not expected to arrive with its own data reduction pipeline. There was also the question of Michelle, which would operate as both an imager and spectrometer and would need various data reduction packages able to work reliably in the tough thermal-infrared regime with its high data rates and the need to nod, chop and perhaps mosaic images. Gillian Wright says she was '*Absolutely totally convinced that in the long-term to get the best out of Michelle UKIRT would need some way of doing the 10 micron work on the best nights. Plus we saw Michelle as being like CSG4 in that it would be so new technically that to get publications out quickly would mean providing observers with publication quality data in real or almost real time*'.

Alan led a number of discussions with his software group (which included Phil Daly, Nick Rees, Frossie Economou and [later] Maren Purves) and with some of the UKIRT support scientists. There was a consensus that there was a need to improve the efficiency and ease of observing at UKIRT and so to maintain the scientific publication rate, which had been given such a boost by the data reduction packages associated with CGS4 and IRCAM. The idea was to make it simpler to prepare and



J.K. Davies, *The Life Story of an Infrared Telescope*, Springer Praxis Books,
DOI 10.1007/978-3-319-23579-0_13

carry out observations, to give observers immediate feedback on their observing, to reduce the time wasted at night and to encourage speedy publication by producing quality reduced data at the telescope. There was also a strategic aim to have software that could accommodate queue, or flexible, scheduling in the future. Without this there was a danger that UKIRT would become scientifically uncompetitive in the rapidly approaching era of 8 m telescopes. Alan and Gillian wrote up these ideas in 1996 for a conference on twenty-first century observing strategies in a paper called. "Queue Scheduled Observing at UKIRT – Evolution not Revolution". These were bold goals and meeting them would turn out to be a major effort, eventually replacing all of the software used by observers and providing a standard software interface for all UKIRT instruments. Luckily the project had the full backing of JAC director Ian Robson, who could see it being crucial to the long-term efficiency and effectiveness of UKIRT, and who was looking forward to the time when queue-scheduling would be undertaken.

Nick remembers a meeting in the JAC conference room where the software group thought about what to do with the existing CGS4DR package. They all recognised that the high quality reduced data provided by CGS4DR was a key element in the scientific productivity of this instrument, and that this was a valuable lesson for future instruments, especially for Michelle. Although some patching of existing software might be good enough, the group felt this was not strategically sound if UKIRT was to remain a front-line telescope when the larger, infrared optimised, Gemini North telescope opened a few years hence. With three new instruments (UFTI, Michelle and UIST) expected in quick succession, flexibility was also going to be important. Phil Daly recalls that he had written a memo which described moving away from monolithic programs like CGS4DR and making the astronomer use data products that were '*Instrument signature removed*' rather than raw data. He thought that the way to do this was to have a new pipeline using the IRAF software package to process the data in a known way. Part of the discussion was over whether scripting would be efficient enough when compared to the direct programming of CGS4DR. Frossie says that she '*Had a particular bee in my bonnet about "easily modifiable" as my previous task had been to insert some extra code in CGS4DR to do a correction for a stuck grating or something. CGS4DR was awesome in many ways, but just inserting a step in the process was no easy task, and I really wanted to do something about that, since instruments, strangely, never stick to the plan and behave how they are meant to*'.

At the same time as these ideas began to come together, the computing infrastructure at the JAC was being replaced, moving from DEC VAX's to Sun workstations running the Unix operating system. This would require re-coding or replacing some existing packages that would no longer be supported under the VAX computers' VMS operating system. There were personnel changes too, Colin Aspin, author of the existing IRCAM reduction software had left, and Gillian Wright and Alan Bridger were about to return to Edinburgh, taking with them years of operational experience. At the same time work was going on in Edinburgh developing software for the JCMT and Gemini telescopes, so three strands were converging in a way which offered potential benefits if they could somehow be combined.

Alan Bridger left Hawaii in late June 1996 but, before he had even left Nick Rees had drafted, and Tom Geballe sent, a message to Malcolm Stewart, the head of software at ROE. The message asked whether the first thing Alan could work on once he returned was to convert the existing *UKIRT_PREP* software to Unix. This request fell on largely deaf ears but a need to port *UKIRT_PREP* was a part of the puzzle they were trying to solve. In November 1996, after a period working on the telescope upgrades project, Nick returned to the wider UKIRT software issues, holding a series of '*Whither UKIRT data reduction software*' meetings in December. To Nick an important thing was that Gillian, with many years of experience with CGS4 was still at JAC. Gillian had been thinking along similar lines for some time and so worked with Nick, and with Alan Bridger back at ROE, to formulate plans and priorities. Nick has a strong memory of '*Trying out MS PowerPoint for the first time and putting up some completely contentless sets of printed slides while Gillian had some scruffy scrawled felt pen slides full of content. I know which ones I liked best*'.

Gillian returned to the ROE in January 1997 which created an opportunity to set up a proper project. Nick believed that to give UKIRT the right long-term, flexible infrastructure and have it be backwards compatible meant it had to be implemented by someone deeply knowledgeable about UKIRT operations and its existing software, and who had the vision to do something. So by the time Gillian returned to Edinburgh she had a remit from Nick and Ian Robson to consult with Alan Bridger to see what could be done with the small amount of effort likely to be available and to try and get some key people interested. Between then and mid-March there was quite a bit of behind the scenes chat between Hawaii and Edinburgh to develop a proposal for a whole new set of software and to estimate the effort required to do it. An outline scheme was sent officially to Malcolm Stewart and Donald Pettie at ROE on the 17th of March 1997 but, while the document was still being drafted, Nick wrote to Gillian and Alan '*The form will be deliberately very rough, since I want to ensure public ROE involvement (and, hopefully, enthusiasm) in the project definition phase. I have seen all too often someone present a very well thought out and defined plan only for it to fall on deaf or negative ears since the people who will be responsible for implementing it have an understandable lack of enthusiasm since it was not their idea*'. This strategy must have worked because this time the response was more receptive. Nick and Frossie came to Edinburgh in early April to flesh out some details of the proposed project. Gillian remembers Nick '*being quite firm about the fact that this was not a project just any person at ROE could take on, but had to be Alan and me because of our recent operations background*'.

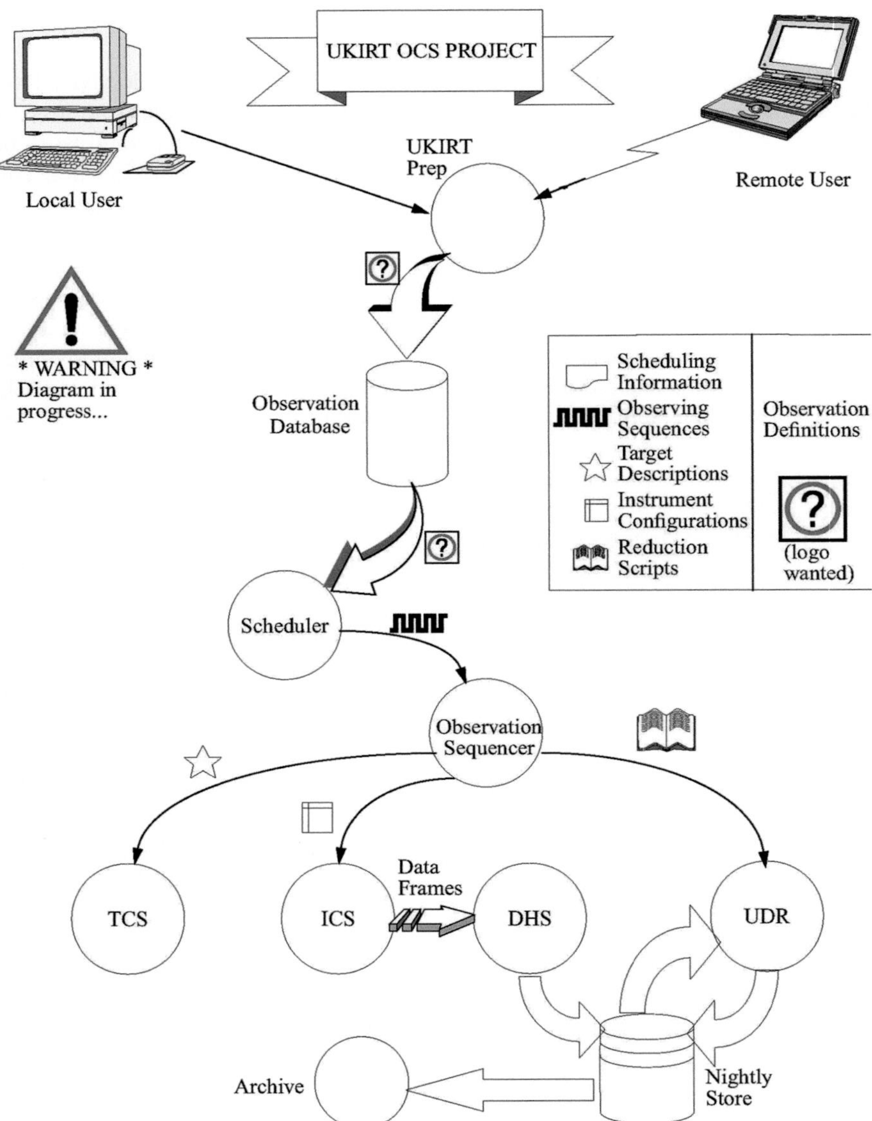


**Fig. 13.1** An early schematic diagram of what would become ORAC (Photo ROE)

## From Design to Reality

The project started in the summer of 1997 and while Frossie, Alan and Gillian worked on the basic design for the new software, Nick sat down with Ian Bryson (then the Michelle project manager) and was shown how to magic about four staff years of software effort out of the ROE system. Some of the effort would come from Michelle, although according to Nick '*The Michelle project had basically put very little thought into providing CGS4DR type software, and when it was raised made the point that it wasn't in the project deliverables and they didn't have the effort for it. Consequently, we had to do this if Michelle was to work*'. Further effort would be found from other projects that would need to interface with the new systems and some of it would require a new programmer willing to work at ROE and then transfer to Hawaii to support the new software once it became operational. This person could be found by recruiting into a vacancy soon to occur in the JAC software group, a post which could temporarily be relocated to Edinburgh.

The new member of the ORAC team would be Malcolm Currie, who moved from a software position at the Rutherford and Appleton Laboratory at Harwell, near Oxford. This was before the days you could send e-mail from mobile telephones and Frossie remembers '*Walking up Blackford Hill with pneumonia to send him an e-mail to encourage him to apply for the vacancy*'. Malcolm worked on the routines for imaging, and Frossie did the pipeline and spectroscopy routines. Also coming on board in mid-1997 was Andy Adamson, who was then at the University of Central Lancashire in Preston, and who joined as external project scientist. Andy was a regular UKIRT user and also worked in software, so he combined two important attributes in one person. Gillian Wright says that '*Once the project was conceived we needed a proper external project scientist. Frossie, Alan and I put our heads together, and I wrote to Andy. After some questions and careful thought he accepted the role, much to all our delight*'. Brad Cavanagh, from the University of Victoria, also worked extensively on ORAC-DR, first as a student on work experience at JAC and later after he became a member of the software group in 2002.

The overall aim of all this activity was to provide a portable infrastructure that was extensible to all existing and future UKIRT instruments. If this could be done then new instruments would only need to have software specific to that instrument, provided this met the interfaces defined by the telescope, everything would work together. Additional software for flexible scheduling could be added later since all the required "hooks" were to be put in place at this early stage. Originally known as UKIRTDR, the project would eventually be called the Observatory Reduction and Acquisition Control system (ORAC). It was not a co-incidence that this was also the name of the cantankerous computer in the Blake's 7 TV series; like its fictional namesake, ORAC would take over the running of things. The real ORAC would have three main components: a new observation preparation system, an observatory control system and a flexible on-line data reduction system.

The observation preparation tool, ORAC-OT, would replace the venerable *UKIRT_PREP text* editor with a modern graphical interface. *UKIRT_PREP* was popular and had significantly improved efficiency by stopping people spending much of the night on setting up instrument configurations. Indeed, some observers wanted to run *UKIRT_PREP* in the UK, not just in Hawaii. Unfortunately since *UKIRT_PREP* was a stripped out element of the UKIRT instrument control system it was not supportable elsewhere. It was also becoming clear that the increased dependence on guide stars (as a result of the telescope upgrades) meant that adding a "check guide stars" capability to *UKIRT_PREP* would be needed to further improve efficiency.

This new observing preparation software was based on the Observing Tool being developed for the Gemini telescope. Although this meant some stylistic compromises, there were many advantages of re-using or modifying Gemini software packages, most importantly that more could be done within the limited ORAC budget. Alan Bridger led the work to make the Gemini Observing Tool able to support more than one observatory, changing its basis significantly and fixing a lot of bugs, thus giving something back to Gemini in the process. Adrian Russell, director of the UKATC at the time, helped with the politics of getting Gemini to agree that UKIRT could share the OT development. The collaboration with Gemini was not, however, always smooth. The Gemini team had their own agenda and were not always willing to listen to, or take account of, the views of others. A few years after the project was underway a small workshop was held in Hilo to investigate the extent to which the JAC could derive mutual benefit by cooperating with Gemini in the development of high-level software and relations became very strained. The “we know best” attitude of some of the Gemini staff so outraged the UKATC team that at one point they were on the verge of walking out and heading straight for the airport. Nick Rees and Andy Adamson managed to calm things down that evening, but the next day Andy was so pre-occupied that while driving into work he went into a 35 mph zone at twice the speed limit and straight into the arms of a police radar trap.

One of many new features in ORAC-OT was one that allowed an observer to enter the position of a target and then see an image from an on-line digital sky survey database showing what the surrounding region of sky would look like. This allowed guide stars to be selected and observation parameters (such as image mosaic patterns and spectrometer slit angles) to be tuned to avoid (or include) other nearby objects long before actually getting to the telescope and losing precious observing time setting up the observation. The new interface was very different, and very much more powerful, than *UKIRT_PREP*, but it would require some mental adjustments by astronomers and was not without its idiosyncrasies. Thor Wold wrote in his newsletter column ‘*I have a feeling that observers, who will have to show up with their entire observing plan (down to guide stars), will experience some growing pains over this. To whit: thus far, not one soul that I have asked has been able to guess correctly what the dog icon stands for in the Observing Tool*’.[1]

The Observatory Control System (OCS) controlled the interface between the telescope and its increasingly more complicated instruments. It was designed such that it could be retro-fitted to existing instruments such as IRCAM and CGS4.

ORAC-DR was the data reduction part of the project. With the ground-breaking work UKIRT and ROE software staff had done with CGS4DR by producing virtually fully reduced data automatically, it was agreed that the ORAC reduction

[1] It was to “fetch” an observing programme from the database.

pipeline should aim to provide science-quality data and be completely data-driven. It was also intended to extend automatic data reduction to imaging instruments like IRCAM and its successors. This meant that all information about the data and its purpose needed to be encoded in the data headers of each observation. To provide both increased flexibility and reduce support needs, the ORAC-DR system was designed to use a pipeline to control existing data reduction packages that were already available elsewhere. These packages would be integrated into building blocks that could be re-arranged and controlled by some scientist-friendly "recipes" that could specify on a human-readable level what was being done to the data. For example, there was a recipe that said REMOVE BIAS and FLAT FIELD. To do this there was a common block that knew how a flat-field correction should actually be applied, and an algorithm engine from elsewhere would be called to perform, for example, a division of one frame by another.

The decision was taken to use the UK Starlink software package because one of the things that was vital to ORAC was for it to have a robust pipeline. The other obvious candidate was IRAF, but at the time IRAF did not have a suitable programming interface. Frossie put it this way '*The only way to tell whether something had succeeded or not was to parse some output and look for the word Error – which was never spelled consistently. For a non-interactive pipeline ORAC needed to be able to tell the difference between "I failed to fit your data because there aren't enough sources in it" and "I fell over because you are out of disk space"*'. Starlink had not only a proper programming interface with real status returns but also several other advantages, for example proper variance handling across all applications, a history mechanism that recorded all operations on the data, error propagation and so on. Gillian insisted that the IRAF option was looked at closely just to be sure nothing was missed, but there was no interface suitable for scripting. The Space Telescope Science Institute eventually implemented a "scriptable" interface to IRAF (called PyRAF) but it was not available when it was needed by ORAC-DR. This decision, sound though it was at the time, was possibly the largest single reason other US observatories did not adopt ORAC-DR, even ones like Gemini with which there could have been synergy.

Not long after the project started Tim Jenness (who had joined the JCMT software team from the MRAO in Cambridge) and Frossie started spending a lot of time as she puts it "*Talking in the pub*". Tim was starting to work on data reduction for the SCUBA instrument on the JCMT and he also wanted something that was flexible and portable. At the time UKIRT and JCMT had completely different software groups with very little shared infrastructure, but Tim and Frossie were becoming friends and had a clear vision of the engineering (and personal) advantages of a common approach. What better way to ensure flexibility than to design a pipeline that could reduce data from both CGS4 on UKIRT and SCUBA on JCMT, two totally different instruments from different telescopes with different conventions and operational models that probed different physics? Looking back Frossie says '*This decision paid off in many ways and continues to do so today. First, it really did work – the pipeline was so flexible, that later it got adopted by a totally different facility* (*the AAT for IRIS-2*)*. Secondly, it brought Tim's shall we* *say more rigorous approach to software engineering on board. Thirdly, it created a common product between UKIRT and JCMT that I think was a not insignificant factor in the decision to merge the UKIRT and JCMT software groups after Richard Prestage's departure for NRAO, with Nick taking over the unified group*'.

The first commit to the ORAC-DR source code repository, essentially the laying of the first brick in a building, was by Frossie on Tuesday the 3rd of February 1998 at 14:51 HST. The intention was that the first functional versions of the preparation system and data reduction pipeline would be delivered to UKIRT in the summer of 1998 to be in place for the arrival of the new imager, UFTI. Later, Michelle would be the first instrument at UKIRT for which the user-interface was the full ORAC system. In the end it didn't quite work out this way. ORAC-DR for the JCMT was ready first and went into use at JCMT on the 25th of July 1998. ORAC was installed at UKIRT in October 1999, in time for the first trial of the new Telescope Control System. After so much effort by the team Andy Adamson described seeing it all come together as '*nothing short of an emotional experience*'. ORAC was initially commissioned with UFTI, replacing the interim software with which UFTI had been commissioned.

## Millennium Bug

The much heralded end of the millennium came and went at UKIRT without incident. In the months leading up to the 31st of December 1999 the telescope and instrument software had been carefully checked for any risks associated with the so-called millennium bug and while the software team was convinced they were safe, no observing was scheduled on the big night. However there was to be a sting in the bug's tail for, unlike most century years, the year 2000 was a leap year. Tom Geballe was observing on the 28th of February when at midnight the ORAC-DR data reduction system suddenly crashed and could not be restarted. It seems that while ORAC knew about the leap year the telescope software did not and so for the rest of night the observers were obliged to take data without any idea of how good it was. The only consolation for Tom was the he received an assurance that he would not be scheduled on the next century leap day, the 28th of February 2400!

## Big Bang

Due to the obvious success of the new system it was soon decided to retro-fit it to CGS4 and IRCAM and deploy it in one fell swoop for all active UKIRT instruments – a scheme that the project team called the ORAC Big Bang. It went well and the ORAC software was integrated with all UKIRT instruments in May 2000. After some testing by local staff for both scientific and engineering work, ORAC was used for all observing programmes after the 1st of August 2000 and has been ever since. Thor's fears turned out to be (mostly) unfounded since by August he was able to write that '*Things are progressing nicely and it seems the visitors are catching on quickly*'.

At about the same time the data reduction pipeline (ORAC-DR) was released to the Starlink community to allow UKIRT and JCMT observers to re-reduce their data at their home institutions if they so wished. Completing the revolution was a further new piece of software called WORF. Another SF related acronym this was contrived to mean "WWW Observing Remotely Facility" and, now that new computer networks could cope with the data volumes from array instruments, allowed remote eavesdropping of the data being taken by UFTI and CGS4. With an 11 hour time difference between Hawaii and the UK, this made it possible for UK astronomers to watch their data being taken in what were virtually office hours at their home institution. The intention was to allow co-investigators to assess the quality of the observations as they were taken without imposing additional work on the observer at the telescope.

ORAC was delivered on time and on budget, a tribute to what a small dedicated team that have worked well together for a long time can achieve and which to Gillian was '*Just typically UKIRT. There was a real creative energy between the main players, and we had a good laugh*'. More formally another paper, "ORAC: A Modern Observing System for UKIRT," outlined the post-delivery system and emphasised the cheapness and benefits from the re-use of software. Alan remembers a comment from Bob Hanisch (a senior scientist at the Space Telescope Science Institute) along the lines of "*We talk about re-using software a lot – you've actually done it*!"

## ORAC's Legacy

Over the next few years ORAC was further developed by the software team in Hawaii and eventually went on to fulfil its mission to reduce data not only from all future UKIRT instruments but also for those on JCMT with its eventual adoption for ACSIS and then for SCUBA-2. Today it powers the JCMT Science Archive, proving beyond the shadow of a doubt that astronomical data can be reduced to a publication-ready level without direct human interaction. Another consequence of the decision to merge the groups was one of the many factors that enabled UKIRT to go into "minimalist" mode 15 years later, since by that time it had a huge common supported software base with the still funded JCMT.

There were two other lasting legacies of the close co-operation brought on by the ORAC project, Tim and Frossie married in 2004, Alan and Gillian in 2008.

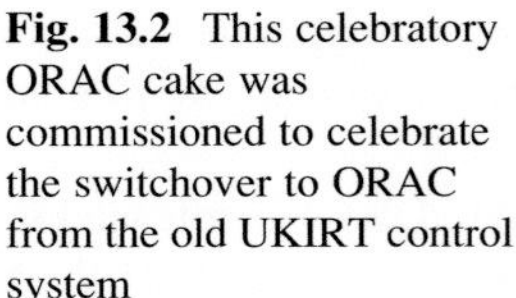

**Fig. 13.2** This celebratory ORAC cake was commissioned to celebrate the switchover to ORAC from the old UKIRT control system

## Staff Changes

While ORAC development had been continuing there were a number of changes amongst the staff in Hawaii. A major departure was in October 1998 when, after 17 years at UKIRT, Tom Geballe left to join Gemini. Amongst the candidates to take his job (no-one could say they were going to replace him) were John Davies and Sandy Leggett from UKIRT itself, Phil James from Liverpool and Andy Adamson from Preston. In the end the job went to Andy, who arrived as the new Astronomer In Charge in October 1998. John admits to being bitterly disappointed by his failure. '*It had been a long road from a desk in flight test at Warton, via Leicester, RAL, Birmingham and ROE to Hawaii. I'd been working on UKIRT, apprenticing for the job, for a decade. It was heartbreaking to fall at the last fence. I was devastated. Still, as someone said to me, you can't blame Andy, it wasn't his fault, so I didn't and I think we got on just fine afterwards. To be fair, he did turn out to be a very good choice*'.

In the summer of 1999, Antonio Chrysostomou, the support scientist responsible for polarimetry and wavefront sensing returned to Hertfordshire after two and a half years in Hawaii, creating a vacancy in the support scientist group. After a short interview and a very long debate amongst the selection panel, Chris Davis was promoted to fill the vacancy; he would remain at UKIRT in various positions until 2009. After narrowly losing out to Chris for the support scientist position, Stuart Ryder left to join the Anglo-Australian Observatory in October 1999. The resulting two vacancies were soon filled, Olga Kuhn arrived in October and Watson Varricatt in December to occupy the second generation of split TSS/PDRA positions.

About the same time Paul Hirst arrived as a PhD student on PPARC studentship for 2 months during which, as well as hunting for data for his own project, he helped with the development of a much needed archive of UKIRT data and got some observing practice at the summit. This experience proved useful in setting some career goals and in March 2000 he returned to take a position as a support astronomer. He recalls '*When I interviewed for the position, we got the telescope tour, which Tom Kerr led. We walked into the dome and Mike Wagner was working on CGS4. Just at the moment that Tom led us interviewees into the dome a distinct cloud of blue smoke rose out of the electronics that Mike was working on. Everyone played it cool and I understand Mike fixed it up before nightfall*'.

**Fig. 13.3** The UKIRT TSS team together on the summit for some first aid training. L-R: Watson Varricatt, Tim Carroll, John Davies (the TSS manager at the time), Olga Kuhn and Thor Wold. The shift patterns meant that getting all four TSS together was a very rare event (Photo John Davies)

Mike remembers the event well. He had been trying to track down a fault in the ALICE electronics rack and had exhausted every idea he could think of except replacing the backplane. Thinking it would be a simple task; he removed all the cards and then the backplane, carefully noting what came from where so it would go back together just as it came apart. However, unbeknownst to him, a plastic insulating washer had been left out during the original assembly. Mike duplicated the mistake by reassembling it '*Just the way I took it apart*' but evidently he tightened the naked screw a little more than when it had been originally assembled, grounding the rail to the frame. Just as Tom Kerr walked into the dome to show the group of support astronomer candidates the telescope, he plugged in the ALICE crate and promptly let the smoke out. He describes it as '*My most stressful day ever, removing that burned backplane and putting back the old one- this time with one more plastic washer in than came out*'.

## Another Red Alert

After his promotion to Support Scientist, Chris Davis encountered a medical emergency reminiscent of some of the red alerts in the past. In 2001 he was supporting a group of visiting Korean Astronomers who were studying the Galactic Centre and the team had brought along a student for his first observing run. According to Chris '*The student was a young, healthy, fit-looking chap, so was of course doomed. He seemed ok on night one and a little under the weather on night two. As soon as we arrived at the telescope on night three I knew he couldn't stay. He was as white as a sheet and had a worried look on his face. I left the TSS with the observer and started back down to Hale Pohaku. Before we even reached the dirt road the student was losing feeling in his fingers and face. He was starting to panic so I hit the gas. By the time we reached Hale Pohaku the student had passed out. I dashed in, grabbed the oxygen and told one of the cooks to ring an ambulance. Thirty minutes later I rendezvoused with the ambulance on the saddle road, the student still unconscious though thankfully breathing*'. The casualty spent a night in Hilo Medical Center and was as right as rain a few days later. This experience of the young and fit being the first to fall was not unique to Chris. John Davies remembers a visiting observer who looked like he could do 15 rounds with the World heavyweight champion but who vanished below stairs during the night. When he had not appeared after a decent interval John went down and found him mopping the floor, having been sick as dog and then recovered enough to get back onto his feet and clean up.

# Chapter 14
# Michelle

Although UKIRT had been designed as an infrared telescope, and in its early days had operated a number of mid-infrared instruments including UKT7, UKT8, IRASFU, and CGS3, in later years it proved remarkably difficult to equip it with array based instruments operating in the 10 and 20 μm wavelength regions. The first use of a mid-infrared array camera was one developed at the NASA Goddard Spaceflight Center by Daniel Gezari and which was brought to UKIRT in July 1989 for polarimetric observations of the Galactic centre. At ROE early attempts to produce a thermal infrared array instrument focussed on collaboration with Reinhard Genzel's group at the Max Planck Institute for Extraterrestrial Physics (MPE) in Garching. After some preliminary discussions Terry Lee and a few others met with the Germans in the Heathrow Business Centre in early 1989 to set up a joint ROE/MPE project to build a low-cost instrument that would provide experience of operating array instruments in the thermal infrared regime. It was called MIRACLE, the Mid IR Camera of Least Effort.

Terry Lee begged the now decommissioned UKT7 cryostat from UKIRT and John Storey from the University of New South Wales, who was spending six months on sabbatical at the MPE, came to Edinburgh and took it to Germany. He worked with Murray Cameron, also at MPE, to combine components of UKT7 with new optics and an array to make a mid-infrared camera. The detector was a 62 × 58 Si: As photoconductor array bonded to the same multiplexer used with the InSb arrays in IRCAM and other instruments. The result was an array operating in the 7–22 μm range which could, in theory, be read out by the same electronics as UKIRT's existing near IR instruments. Terry went to Garching and, with Murray Cameron, bench tested the system with the MPE InSb array electronics. These tests revealed some problems so the MPE built a dedicated set of array electronics and after more lab testing, MIRACLE was taken to UKIRT for an observing run in November 1991. The split four night commissioning run (on the 1st, 2nd, 5th and 6th of November 1991) was successful and, as well as engineering tests, included scientific observations of the Orion KL complex and the Seyfert 2 galaxy NGC1068. Looking back on the project, Terry says '*For me it was fun, a chance to do*



J.K. Davies, *The Life Story of an Infrared Telescope*, Springer Praxis Books,
DOI 10.1007/978-3-319-23579-0_14

*something hands on, which I was unable to do in Edinburgh.* MIRACLE did not however return to UKIRT. Although comparable to Dan Gezari's camera, its overall sensitivity was no higher and it would clearly not be able to compete with mid-infrared cameras already being built around new, larger and more sensitive detectors.

With 128 × 128 pixel mid-infrared detectors becoming available it was hoped that there would be an opportunity to build on the design experience developed with CGS4 to provide UKIRT with a mid-infrared array spectrograph. The scientific case was originally made by Dave Aitken, but Pat Roche would eventually act as the external project scientist, representing the community's interests for the lifetime of the project. A key part of the justification was that the instrument would be able to provide ground-based follow-up for the European Infrared Space Observatory satellite (ISO), then due for launch in 1993.

Some early ideas were set out by 1990 and ROE's Alistair Glasse was soon named as the project scientist. Although due to shortages of effort in Edinburgh progress was slow, by the autumn of 1991 work had begun at ROE on a design study for such an instrument. The intention was to submit a bid the following year for approval to build the spectrograph as part of the ROE's ongoing development programme. However the funding landscape was changing and the University of Oxford was interested in building up its own group able to build infrared instruments. One option which was discussed was that Pat Roche at Oxford might lead a team who would build a camera operating at 10 μm. This could be developed and tested in parallel to the work being done at ROE on the larger and more complicated spectrograph. The idea was that characterisation of the detector array, its electronics and the required closed cycle coolers could be done early on at Oxford and this experience could then be fed into the ROE programme to help in the developments there. The Oxford work would be funded by a direct grant to the university, but could be assisted by the ROE providing a spare IRCAM cryostat and design drawings so that a set of ALICE readout electronics could be built Oxford. It sounded fine in theory, although there would clearly be a need for careful co-ordination between the two groups and the resolution of some high level policy issues. There were also some purely technical matters; for example could the IRCAM cooler be adapted to work at the very low temperatures required by at 10 μm camera? However the wider UKIRT community was clearly in favour of a high resolution spectrograph and in the end the plan to build a camera at Oxford did not go ahead (although the university would ultimately succeed in building a near infrared camera for the William Herschel Telescope and later one for UKIRT as well).

So the efforts at ROE continued and approval to go ahead with the project was sought in the spring of 1992 with delivery anticipated in 1996. The instrument was known briefly as CGS5 (and also LIRGAS) but the name was soon changed, the name Michelle being chosen by Alistair Glasse from the entries in a competition published in the Observatory newsletter. The winner, a relative of an ROE secretary living in Malmesbury, Wiltshire, duly received a hand delivered bottle of malt whisky as his prize on a dark and rainy evening in the winter of 1992. A shortage of key staff, which was to become a recurring problem, slowed the early stages of the project and soon delayed its originally planned delivery by almost a year. In the meantime, further attempts were made to encourage visiting groups to bring mid-infrared instruments to UKIRT and to offer them for collaborative projects with UK astronomers.

MIRAC, the Mid Infrared Array Camera built by "Bill" Hoffmann, came to UKIRT a number of times but the MIRAC team were quite robust in demanding publication rights in return for the use of their instrument and this discouraged at least some UK astronomers from going down this route. A key member of the MIRAC team also had an uncanny knack of creating tension between himself and at least one of the UKIRT telescope operators over his choice of music to be played in the control room during observing. As a result at least one shift was carried out in near total silence after attempts to find a musical compromise "crashed and burned" on the subject of percussion.

MICS, (Mid-Infrared Camera Spectrometer) was a prototype of an instrument to be built by the National Astronomical Observatory of Japan for the planned 8 m Subaru telescope and could operate as a camera and as a spectrometer in the 7.6–13.6 μm range. It was equipped with a 128 × 128 Si::As BIB array. The camera had a 49 × 49 arcsecond field of view and the spectrometer a slit with a resolving power of between 77 and 128 depending on wavelength. MICS was successfully tested on UKIRT in 1997 and was offered to the UKIRT community in 1998. Unlike the arrangements for MIRAC, there were no requirements for approval by, or collaboration with, the MICS team although voluntary collaboration would clearly be an advantage to everyone. MIRAS, an Australian mid-infrared camera equipped with a polarimeter was a visitor for a few runs in the mid-1990s. In addition to these, as part of the upgrades programme collaboration, UKIRT hosted a German built mid-infrared camera called MAX on numerous occasions, including a series of runs between 1997 and 2000.

None of these collaborative efforts were a satisfactory substitute for a UKIRT-owned instrument, but progress on Michelle was slow throughout the early 1990s because of the diversion of key technical staff at ROE to troubleshoot other projects, notably the much delayed SCUBA instrument for the JCMT. These problems were compounded by a recruitment freeze, which meant that additional effort could not be found from outside the Observatory. A further factor was the negotiation of a contribution towards the development cost of Michelle from the Gemini project. The contribution came after an agreement that the instrument would have the ability to operate on the Gemini North telescope and the inclusion of an imaging mode which would use the same filters as the spectrometer but which would have a dedicated optical path ensuring high transmission when Michelle was used as a camera. The intention was that once Michelle had been commissioned it would be shared on a 50:50 basis with Gemini. This would require interchangeable fore-optics to allow Michelle to switch from UKIRT's f/35 to Gemini's f/16 focal ratio.

Enough progress was made for various preliminary design reviews to be held in 1994, at which time delivery to Hawaii was expected in April 1997, but by 1995

Michelle had fallen even further behind schedule with delivery now aimed at October 1997 and commissioning in March-April 1998. Accordingly PPARC, in conjunction with Ian Robson at the JAC, decided on a review to examine progress, reaffirm the original scientific aims and timeliness of the project and to assess the continuing relevance of Michelle to Gemini North. UKIRT Board chairman Bob Carswell was not convinced about the need for the review describing it as *'A substitute for getting the job done in the best Yes Minister tradition'* and asking *'Why, please can't the observatories just be left to get on with job?'* His view was that the UKIRT board was happy with the project and that the more committees became involved, the greater would be the consequent muddle, wasted effort and delay. Nonetheless, the review went ahead with a panel of three chaired by well respected infrared astronomer Professor Peter Clegg. The panel met at ROE on 10 April 1995 and were expected to deliver their report in a few weeks. In the meantime Terry Lee was advised to *'Minimise any new commitments pending the final outcome'* although it was stressed that this cautious approach should not be read as revealing a negative feeling towards Michelle.

In the end the panel's report, sent to the ROE Director in August, supported Carswell's opinion that the main problems were not so much technical, although some technical issues did need solving, but stemmed from a failure of the ROE to provide the appropriate staff on the agreed schedule. It also noted that further delays in providing this effort would cause more slips in the schedule and reduce the time available for the instrument to be exploited by UK astronomers on UKIRT before the transfers to Gemini began. More importantly, they recognised that although Michelle could not now be ready in time for support of the ISO mission, if delivered to the agreed specification it would be a world-class instrument with unique capabilities. Michelle had passed another milestone, or dodged another bullet, but it would not be the last. Hopes of an early delivery foundered on a set of problems mainly centred on the grating exchange mechanism. This was a large and heavy drum holding five interchangeable diffraction gratings, which had to be accurately and stably rotated about two axes at cryogenic (20 K) temperatures. This tough engineering challenge was eventually solved by the ROE engineers, but only after over a year of delay.

As presaged earlier, the continuing delays meant that Michelle was the subject of yet more reviews in the early summer of 2000 and the project found itself working to a formal cost limit set by PPARC. By now Ian Robson had become increasingly disappointed by the ROE's failure to deliver Michelle and he made a point of attending a number of reviews, which did not increase his confidence in a speedy resolution to the problems. However, there was no doubt that constructing Michelle was a huge technical challenge and he still believes that *'Michelle must be one of the most complex cryo-mechanical instruments ever constructed, even more challenging when one considers the sub-10 K operation. Nevertheless, the ability to maintain schedule and cost was extremely disappointing and definitely hurt the scientific productivity of UKIRT and as it turned out, Michelle itself'.*

Against this background, acceptance tests were carried out in Edinburgh in May 2001, with the grating mechanism now working properly and the instrument generally performing well. In view of the delays, and increasing pressure from higher levels that if Michelle did not go to UKIRT soon it might never go at all, a decision was taken to ship it and deal with any remaining minor issues in Hawaii. With Michelle went Alistair Glasse and his family on secondment for the commissioning period.

Michelle finally arrived in Hilo in June 2001 and was successfully assembled and tested in the Gemini building adjacent to the JAC (where there was more space) before being shipped to the summit in late July. First light was on the 22nd of August when just a single night was allocated almost immediately after the instrument had been mounted on the telescope. In spite of the fact that all data taking and instrument control was being carried out using an engineering interface, the major functions of the instrument were all successfully demonstrated. There followed a seven week commissioning period. In local instrument scientist Tom Kerr's opinion this went *'fantastically well'* although it did highlight some deficiencies in the data reduction software, which had resulted from inadequate testing, itself a consequence of the cost-capped budget and resulting compressed delivery timescale. Tom and Alistair Glasse would take over the telescope for 3 or 4 hours after the regular night of observing finished at around 6 o'clock and work on into the morning, stopping before the Sun got too high to safely continue. They were able to commission every observing mode, imaging, spectroscopy, imaging polarimetry and spectro-polarimetry and it all went well.

The first regular observing run with Michelle started on the 15th of October 2001 and was to do spectropolarimetry, one of the most complicated modes and the last thing you might ever want to start with. It went quite well except for a problem with the top-end of the telescope, which hampered its ability to chop, but that was not the fault of the instrument. Tom recalls *'we were never quite sure where it was chopping as it would just randomly change every so often and we had to reconfigure things on the fly, which involved a lot of calculations at 14,000 ft which was kind of interesting'*.

In its first 5 months of operations, supported by Alistair Glasse and Tom Kerr, Michelle worked extremely well for such a complex instrument and showed clearly the leap in performance at thermal infrared wavelengths that it had brought to UKIRT. The only notable problem was the abandonment of a few runs after the cryostat's potassium bromide front window become opaque because of water damage. The system for flushing the window with dry-air had been overwhelmed by a major storm on the summit, so Michelle was warmed up and the window replaced, a task taking several weeks. The same problem occurred again soon afterwards when the air purge ran out, and so remedial action was required. This involved fitting an additional sacrificial thin window in front of the main cryostat window, which could be swapped quickly without a warm-up. Once this fix was in place there were no further instances of water damage. Michelle remained at UKIRT until the end of 2002 when it was transferred to Gemini North for on-telescope commissioning in early 2003.

**Fig. 14.1** Michelle mounted on UKIRT (Photo Neil McBride)

The plan had originally been to swap Michelle between Gemini and UKIRT several times but despite the objections of Ian Robson a final decision to commit Michelle to Gemini for the long term was made by PPARC in Autumn 2003, with the last runs on UKIRT taking between February and April 2004. It might have been possible to continue to operate the instrument for periods at UKIRT, but when the wide field camera arrived and blocked the Cassegrain focus it would have meant that at most the instrument would only have been available for a few months a year. Supporting this would have been problematic and so UKIRT loaned Michelle to Gemini indefinitely.

Sadly, the potential 50-fold gain in observing speed that should have come from moving Michelle to Gemini seldom materialised as Gemini's chopping system was unable to guide in both beams when driven by Michelle's chop signal. Alastair notes that '*This was a fundamental problem with the telescope design, which arose because the [Gemini] secondary was provided as a black box unit*'. After some of these issues had been resolved, Michelle was operated with reasonable success on Gemini and this is evidenced by a mail from Doug Simons to then Gemini Director Matt Mountain which was ultimately forwarded to Alistair Glasse by one of the Gemini team and which said '*Eric (Becklin) stopped by my office, mentioned that he detected a ~1 milliJansky source last night with MICHELLE, and said MICHELLE is the best MIR instrument he's ever seen. As his former student, I assure you words of praise like that don't come from Eric very often*'.

The sensitivity noted by Becklin was in large part due to the exceptionally low emissivity of the combination of Michelle and the Gemini North telescope, the latter using protected silver mirror coatings for this reason. Despite this capability Michelle was never in high demand on Gemini, sometimes being allocated only a few hours of observing time in a six month semester. The hoped for surge in demand for follow-up of infrared observations made by the NASA infrared satellite Spitzer never materialized, as Spitzer was often used to observe targets that were simply too faint to detect from the ground. With UKIRT's and Gemini's declining budgets, only modest demand from a small mid-infrared community, and resources becoming less able to support such a complex instrument, Michelle never reached its full potential. In 2013 Ian Robson reflected on what he called *'a huge opportunity loss for the UK and especially UKIRT which aptly demonstrated that sometimes delays conspire against eventual success. The fact that Michelle never achieved its world class scientific potential is a sad fact, especially for the dedicated team of engineers and scientists in Edinburgh who spent many years overcoming immense technical challenges. That the Gemini Telescope could not cope with the instrument in the early days just added insult to injury'.*

# Chapter 15
# Departures

## The UK Joins ESO

By late 2000 the UK was in discussions to join the European Southern Observatory, ESO. Despite the injection of extra government funds for the initial investment, in the longer term the financial commitment would require considerable savings from the rest of the UK's ground-based astronomy programme. Accordingly, Ian Robson, and the other UK observatory directors, were required to put forward various cost saving options that might be implemented over the following five years. In response, Ian and Andy Adamson developed three savings models for UKIRT, each with very different levels of savings and operational risk. These ranged from moving gradually to wide-field operations with a resulting run down of staff by natural wastage, to a rapid aggressive savings-driven model which reduced telescope operations and support to the bare minimum. The gradual run-down was considered to save insufficient money and the aggressive plan was thought too risky as it would drastically affect UKIRT just as the considerable investments in new instruments should be paying dividends. So a halfway house model with reduced user support and an intermediate level of change was pencilled into the long-term plans. This, and other elements of restructuring in the UK, including a gradual withdrawal from the Anglo-Australian Observatory and the trading of the VISTA telescope to ESO, was eventually accepted allowing the UK to join ESO from the 1st of July 2002.

For the few remaining UK based staff who were on overseas tours from Edinburgh, the cash savings required to pay for the ESO subscription represented a real threat. Sooner or later, and probably sooner, the roof was going to fall in and the links with Edinburgh would be broken completely, leaving them with the choice of going back to no job or resigning from PPARC and staying in Hilo under contract to the University of Hawaii. Ian Robson describes this period as being '*somewhat fraught*' for the UK employed support staff in Hawaii on both telescopes. The underlying problem arose because the original operational concept had been that



J.K. Davies, *The Life Story of an Infrared Telescope*, Springer Praxis Books,
DOI 10.1007/978-3-319-23579-0_15

staff (mostly scientists) would transfer out to Hawaii from Edinburgh for a tour of duty, which would last three years and could perhaps be extended by another three years. However, with the reduction in opportunities back in Edinburgh, plus the satisfaction of working in Hawaii on two world-beating telescopes, the wish to extend these tours indefinitely became almost the norm. This then caused some friction with the locally employed staff, who were doing effectively the same jobs on less favourable terms and conditions. Along with the pressure to reduce costs, Ian Robson was also keen to move back to the tour of duty concept preferably with a minimum complement of UK based staff on tours. The existing UK staff were thus given the option of returning to the UK (on the assumption that a position could be found for them), or transferring to the RCUH local staff payroll. In the end about half the staff transferred to RCUH employment, the rest returned to the UK, either to Edinburgh or Swindon.

In the spring of 2001 Tim Hawarden returned to ROE after 14 years at UKIRT creating a vacancy in the support scientist team that was filled in May by Marc Seigar. Tim was soon followed by John Davies who left in June for Edinburgh to take on a role co-ordinating a large European Commission astronomy infrastructures programme. John was *'Sorry to leave, I really was, but I was caught in a classic middle aged squeeze, my children were nearing university age and my parents were getting older and needed me closer to home. Furthermore, opportunities to get back to jobs in Edinburgh were becoming scarce and, with Andy as Head of UKIRT, there was nowhere for me to go if I stayed. So I bit the bullet and went'*. Jane Buckle arrived in October 2001 and took over as UKIRT scheduler while Douglas Pierce-Price arrived in December to do outreach work combined with some time for personal astronomical research.

However, these planned departures were nothing to the shock felt in the UKIRT community in January 2002 when Sidney Arakaki, the head of JAC Instrumentation Support, died suddenly. Sidney was a key figure at the JAC, with more than two decades of experience, and he had a huge amount of expertise in cryogenics, high vacuum systems and instrumentation. He was also liked and respected by most of the local staff and could be relied upon to identify potential problems to the management team before they became serious. Sidney had been at UKIRT almost since the beginning and his loss left a very big hole in the organisation. Another major change occurred in September 2002 when Ian Robson left the JAC to take up the Deputy Directorship of the UKATC in Edinburgh with the remit to refocus the UKATC's culture more into line with a customer-driven perspective of delivering instruments on time and to budget. Although the majority of Ian's time had been devoted to the JCMT, of which he was director, UKIRT nonetheless owed much of its success in maintaining a viable instrumentation suite to his thorough knowledge of the PPARC system and his excellent relations with the UKIRT Board. Ian was replaced by Gary Davis, then at the University of Saskatchewan and a regular user of JCMT, who took over as director of both JAC and JCMT. At about the same time, Malcolm Currie, who had developed the imaging data reduction recipes for ORAC-DR, returned to the UK after more than three years in Hilo.

## Goodbye to IRCAM

Also leaving in 2002 was IRCAM, which was removed from the telescope for the last time on the 27th of August marking the end of an era for infrared astronomers, not only in the UK, but globally. IRCAM ended its career in an exciting way which took advantage of its capability for fast readout. On August the 20th, its very last night of use, IRCAM was used in an international multi-telescope campaign to observe the occultation of the star P131.1 by Pluto. The goal of the observations was to look for changes in Pluto's (predominantly nitrogen) atmosphere and to measure the planet's size. The observations at UKIRT were made by University of Hawaii astronomer Dave Tholen and Dave Osip from MIT. Sandy Leggett described this multi-telescope, time-critical collaboration as *'A fitting end to IRCAM's many years of superlative service to UKIRT, UK astronomers and infrared astronomy in general'*.

# Chapter 16
# Flexible Scheduling and the OMP

## A New Way of Working

By the late 1990s there was an increasing desire around the world to maximise telescope productivity by moving towards flexible, also called "queue", scheduling. The objective was to reduce the weather lottery of rigidly scheduled nights and to take advantage of specific, but rare, conditions such as good seeing or high transparency at thermal infrared wavelengths. Such schemes were at the heart of the operational models for some of the 8 m telescopes that would soon enter service. However, with the greatly improved image quality now available at UKIRT and the requirement for dry conditions for some observing modes of the Michelle imager-spectrograph, flexible scheduling also offered UKIRT potentially enormous gains. However, although moving to flexible scheduling had been debated for a number of years at conferences and over cups of coffee on cloudy nights, it would be a huge psychological leap for a generation of astronomers used to "owning" specific nights on a telescope. So, to investigate how the protocols, observing tools and human issues of flexible scheduling might work in practice, UKIRT carried out a flexible scheduling experiment in the second half of 1999.

Three pairs of projects that appeared to have complementary weather requirements were identified and linked. These pairs were scheduled contiguously and one observing team from each pair came to the telescope with instructions to "Flex" between the two projects according to the conditions. To simplify the potentially traumatic decision on when to switch projects, detailed rules were laid down on how conditions were to be measured, when the projects were to be changed over and how time would be accounted in the event of bad weather, system failures, etc. All three pairs of runs took place with various combinations of observers, students and staff observers present at the summit. Overall the process did deliver more time on target in optimum conditions than would have been achieved otherwise, but the experiment did teach several important lessons, notably that the observation preparation tool must make due allowance for overheads and that a robust data reduction pipeline with adequate diagnostic tools was essential.



J.K. Davies, *The Life Story of an Infrared Telescope*, Springer Praxis Books,
DOI 10.1007/978-3-319-23579-0_16

This was crucial for the long term when only one team, or perhaps none, would be at the telescope. One of the key objections to flexible scheduling, and to remote observing, was a loss of the ability to do immediate follow up or to adjust the observing plan as necessary. Ian Robson, and others, believed that this could be reduced to a well defined decision tree that could be worked out in advance and followed by any observer or telescope operator provided that the software could reduce the data quickly enough and well enough to enable such a decision tree to be used.

Furthermore, and most importantly, it was clear that to offer routine flexible scheduling without increasing the workload on the visiting observers or the UKIRT staff, software was needed that would automate all stages of the process and keep a clear and correct record of how the observing time was used. ORAC was already close to delivering some of these requirements, and the solution to the other was actually being developed for a telescope only about a kilometre away.

## The Observatory Management Project

As a mainly sub-millimetre astronomer, Ian Robson was well aware that observing at the JCMT was critically dependent on the dryness of the atmosphere so he had been pushing the need for flexible scheduling of the JCMT for some years, often against strong opposition from the user community. However, he had persevered and, along with support from the JCMT management board, had set the train of events in motion that would eventually see flexible, weather-based, queued observing become the norm. The staff had already started on the Observatory Management Project (OMP), which would deliver all that was needed for flexible observing and expanding this to provide UKIRT's needs was an obvious step. Thus in January 2001 the OMP became a JAC-wide project covering both telescopes and which was led by Frossie Economou and Tim Jenness. The project passed its critical design review in July 2001 and the team made some modifications to the plans based on the recommendations from the review. It was however, according to Frossie, '*Stuck with the old name, which irritates me by having no word indicating that it is a software project, and by having no obvious mascot opportunities*' .

The OMP built on the software already developed for ORAC and was designed to handle the entire flexible scheduling process from observation preparation, through making the observations, data reduction and feedback. The process would begin with astronomers specifying their observations exactly as they had done before, but adding information on the required observing conditions and any specific scheduling constraints. All of these potential observations were then submitted to the UKIRT observing pool. At this point what the UKIRT newsletter described as "*magic*" would take place. The OMP had a sophisticated database server that would examine all the observations submitted and extract and store scheduling information from each of them. During each night of observing the OMP would search the database and provide the astronomers at the telescope with a selection of the most appropriate observations based on the weather conditions, instrument availability, scientific ranking, etc. The observers would then be able to choose the project best suited to the conditions, moderated if necessary by their own experience and confidence in their ability to execute the observation successfully. The critical difference from traditional observing would be that the observations to be done would not necessarily be from their own scientific programme, they would be the ones best suited to the moment. This implied that the principal investigator of the project that was being observed had to be informed about the progress of their programme. This feedback was vital to the success of the process because it increased efficiency by encouraging the involvement of the absent scientists. So the OMP informed each principal investigator of what had been done, allowed them to retrieve their data, and then provided an opportunity for them to discuss any issues with the scientific staff to see if the process could be improved.

It sounds simple, but projects always do until you start trying to actually do them. This was to be a complete change to the way UKIRT had operated for two decades and the OMP team had to interact with the entire user community and the staff of the two Hawaii telescopes during the design and implementation phases. Since they were trying to change the way people were used to working they naturally met resistance and Frossie in particular spent a large amount of time seeking out people who were willing to contribute and to support the project in some key areas. As lead technical person Tim Jenness had to cope with a conflicting set of user requirements from the various communities and merge them into a single, coherent, design. The two them then had to distil this design into sub-systems which individual software engineers could produce, motivate their team into delivering something that not everybody wanted, and to do it on time and on budget. During this process Frossie wrote in her blog '*We have a design, we have a review, we tweak the design some more, we have a mountain of dead whiteboard markers, we change our minds a few million times, we drag in anybody willing to help, we have large head-shaped dents on the walls*'.

## Flexible Scheduling of UKIRT Begins

Eventually it all came together and the first and largest OMP release was launched on the JCMT in early August 2002. According to Frossie's blog the software corridor at the JAC was by then '*Full of exhausted software engineers alternatively pulling their hair out, wondering when the last time they went home was and yelling "It's all totally broken!" at each other*'. During March and April 2002, in parallel to the hair pulling and shouting, the OMP had been tested on UKIRT during service observing nights and later in the year the UKIRT management board endorsed a proposal to make the move to full flexible scheduling by the spring of 2003 (i.e. UKIRT Semester 03A). They set out the following principles,

1. Completion of programmes, especially high priority programmes, is the key goal.
2. Programmes requiring the best observing conditions should have the best possible chance of being done in those conditions.

3. Observers who come to the telescope should have a good chance to carry out their own programme.
4. Summit observers must retain a significant degree of discretion in selecting programmes for execution.

This was major change for the UKIRT community. Scientists winning time on the telescope no longer had a free ticket to Hawaii. Instead the time allocation panel granted about half of the successful teams the right to go to the telescope and the remainder, usually those with the least complicated observing programmes, stayed at home. Summit occupancy was not, however, complete control of the telescope, it was the right, or obligation, to execute observations selected from the queue by the OMP. If the conditions were suitable for the project proposed by the summit team, all well and good, they could use their skills to get the best for their programme using whatever specialist knowledge had gained them the right to be there. If not then it was incumbent on them to work just as hard and as diligently to execute a programme belonging to someone else. Altruism and integrity were going to be big factors if this process was to work.

Meanwhile at sea level Frossie Economou and Tim Jenness were jointly named 2002 RCUH employee of the year. The JCMT newsletter article announcing the award was written by Nick Rees and included the following. '*The release of the software spanned a four-month period from July to October 2002 and both Frossie and Tim worked 14 hours a day, 7 days a week during the commissioning period. Frossie worked from 8 am to 10 pm and was present on the mountain to give personal support. Tim worked from 2 pm to 4 am in Hilo using the video conferencing system. The whole process was extended by the run of poor weather we had in early 2002B, and so the support had to be continued way beyond the two weeks initially scheduled until the systems were fully tested. All of this was, of course, extremely draining, and would have taken its toll on lesser individuals. However, throughout the period both Tim and Frossie maintained a positive attitude, dealing personally and calmly with all the new observers' fears and questions, and continuously improving the system as time went on*'. Frossie's blog noted dryly '*I assume the word "calmly" means something else in Australian*'.

The deployment of the OMP on UKIRT in Semester 03A saw the dawn of full flexible scheduling. Everybody involved coped remarkably well with the transition and by the summer of 2003 all UKIRT and JCMT common-user observing was being done under the auspices of the OMP. It soon became clear that the essential goal of flexible scheduling, the preferential completion of the best programmes, was being met in the first semester of operation. However, it also became obvious that the speed of the feedback loop between the observatory and the remote scientists was not always sufficient to guarantee the scientific quality of the data. Resolving this issue was helped when the OMP version of the eavesdropping software WORF became available allowing the absent scientists to eavesdrop on their observations as they were being made.

The OMP activity finished in the autumn of 2003 with Frossie vowing '*Never to get involved again in a project that has the word project in its name, so as to avoid agonising for hours on how to write sentences that read The Observational Management Project project draws to a close*'. Naming aside, the OMP was hugely successful in building a software suite that ran on two completely different telescopes and supported their various, very different, instruments. This increasing commonality between the two JAC operated telescopes also allowed UKIRT to adopt the JCMT observation queue system from January 2004 onwards. This software could execute a large series of observations automatically and, while it was useful for what would turn out to be the last UKIRT observing with Michelle, it would prove vital for the Wide-Field Camera when it arrived.

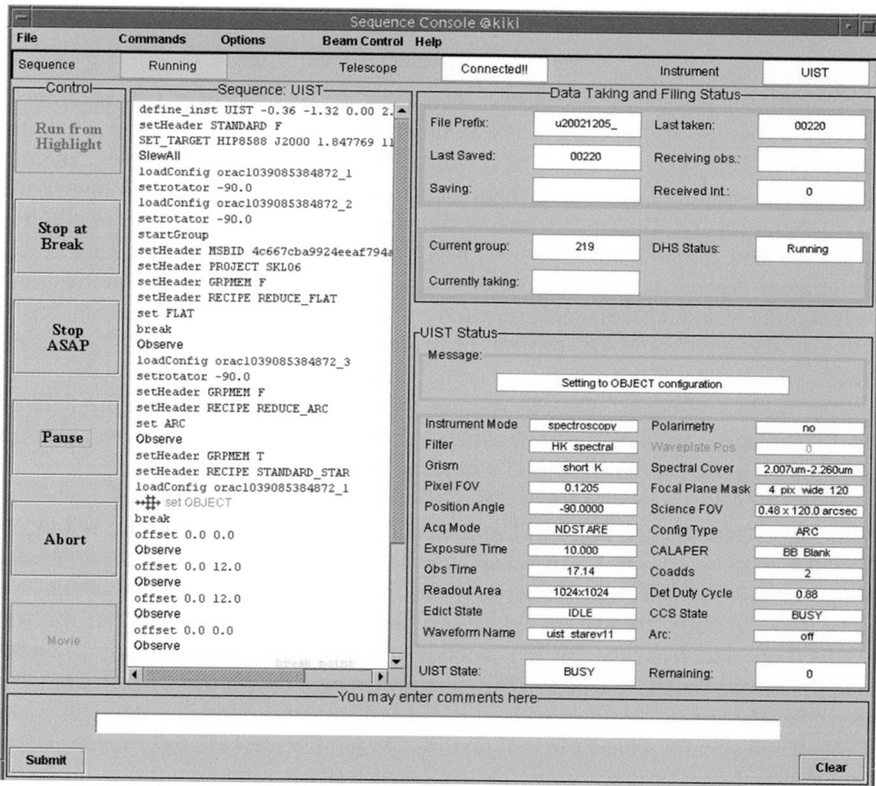


**Fig. 16.1** The "Sequence Console" of the OMP shows the individual steps of an observation, the slew to the source, the configuration of the instrument, and the actual observations interspersed with jitters around the array or slides along the slit (Image JAC)

Flexible scheduling, supported by the tools of the OMP, was a success and would remain the principal operational mode for UKIRT into the future. Ultimately though, however good the tools, flexible scheduling at UKIRT relied on the almost uniquely collaborative UKIRT user community to make it work. Andy Adamson said that '*While there were bumps along the way, my fondest recollection is of an observer who came to the telescope when 95 % of his data had already been taken in the queue, observed other people's projects for 5 nights, and still went away satisfied*'.

Ian Robson had a similar reaction noting that '*The fact that the user community did operate in an altruistic rather than cut-throat fashion was a huge change of culture, and showed how much UKIRT had influenced the individuals; they were prepared to change the habits of a lifetime to continue to see great science being undertaken from a telescope that was definitely under threat at the time. Everyone benefited, which was brilliant, and a very rewarding experience*'.

## *eSTAR*

One small but significant additional advantage of the OMP was that it allowed UKIRT to present itself to the world in much the same way as automated robotic telescopes. This meant that UKIRT, which was much larger than most robotic telescopes, could join networks which reacted to sudden astronomical phenomena like gamma-ray bursts, events which need to be followed up in a matter of minutes to get the best possible scientific return. To catch such events astronomers set up "Intelligent Agent" computer programs which silently roamed the Internet and watched the skies. These agents, created by the "eScience Telescopes for Astronomical Research" (eSTAR) project, provided links between the SWIFT satellite, which was designed specifically to search for sudden astronomical events, and ground based telescopes. The eSTAR network made use of the emerging field of intelligent agent technology to provide crucial autonomous decision making in software to speed-up the issuing of alerts and follow-up requests.

When the OMP was complete it was possible to deploy eSTAR on UKIRT, making it the largest telescope in the world with an automatic response system for following up gamma-ray bursts. The key was that under the OMP all aspects of an observational programme at UKIRT were either software readable or software controllable, so eSTAR could fully specify an immediate follow up observation. When eSTAR received an alert from the SWIFT satellite it would place an observation request directly into the UKIRT queue as a high priority target of opportunity, which would be seen when the human observer next requested an observation. After the data was taken it was automatically reduced in real time by ORAC-DR, and the fully reduced data returned to the eSTAR software. This allowed UKIRT observers to get on target within minutes of the burst occurring, and potentially allowed automatic evaluation and further follow-up of the initial first-look data.

The first successful observation by eSTAR using UKIRT was of gamma-ray burst 050716 where the system responded to the SWIFT position alert and queued an observation at UKIRT just 48 seconds after the initial alert was received, although operational difficulties meant that the observations were not carried out until 56 minutes later. Despite this delay, this showed that the system worked and in future UKIRT would have considerable success following up eSTAR alerts. According to Tim Jenness '*It was the culmination of all the hard work over the previous few years to get a modern infrastructure in place*'.

# Chapter 17
# UIST

With the Observatory Management Project underway the first few years of the new millennium were a busy period at UKIRT. No sooner had Michelle commissioning finished than preparations began for the arrival of another new instrument, the UKIRT Imager SpecTrometer (UIST). This was conceived in the mid-1990s as a way of taking advantage of the newer, larger, InSb infrared arrays that were becoming available. Various concepts for such an instrument were presented to the Optical-Infrared Review Panel and described in UKIRT development plans that followed the outcome of that review. For a while considered as a 1–5 μm spectrometer (CGS6 or CGS6E if equipped with echelle gratings), a multi-purpose near infrared clone of Michelle (called Sichelle) or as a new infrared camera with grisms for spectroscopy (IRCAM-G), a less expensive concept capable of high resolution imaging and moderate resolution spectroscopy was eventually approved. After preliminary and final design reviews in late 1997 and late 1998 respectively, the instrument was built in Edinburgh at the newly established UK Astronomy Technology Centre. Suzie Ramsay, who had done her PhD on CGS4 almost a decade earlier, was the project scientist and Mel Strachan was the project manager. Although UIST was an acronym, it is also the name of a Scottish island and appealed to the Scottish roots of both Suzie herself and the Royal Observatory in Edinburgh. There may have been another reason too, Mel says the team wanted to get away from naming instruments after girls (Michelle, ALICE and Naomi all having been recent ROE projects).

UIST combined imaging and spectroscopy over the 1–5 μm wavelength region and was intended to replace the low resolution spectroscopic capabilities of CGS4 and many of the imaging functions of UFTI and IRCAM. It would also provide UKIRT with improved polarimetric capabilities. Imaging polarimetry was possible with UFTI, but the polarimetry mask was warm and slightly out of focus and the arrangement was not ideal. The even more demanding spectro-polarimetry was possible with CGS4, but its Wollaston prism came before the spectrometer slit which meant that the two flavours of polarised light – the diverging e-beam and



J.K. Davies, *The Life Story of an Infrared Telescope*, Springer Praxis Books,
DOI 10.1007/978-3-319-23579-0_17

o-beam rays – had to be carefully directed down the slit. If this wasn't done precisely then spurious instrumental polarisation was introduced.

UIST was intended to exploit the much improved image quality that was being delivered by UKIRT since the upgrades programme and at its heart was one of the new generation 1024 × 1024 ALADDIN arrays from Raytheon. The plate-scale for imaging was selectable from lenses in a cooled wheel inside the instrument, allowing the observer to respond rapidly to changing observing conditions or to specific scientific requirements. UIST was the first UKIRT instrument to use grisms, instead of slits, for low resolution spectroscopy and it also offered an integral-field mode, which provided 3D spectroscopy in a common-user cryogenic instrument for the first time. Unlike other instruments, in which the integral-field unit was fixed, UIST used a compact device that was mounted on one of the internal wheels and so could be switched in and out of the light path very easily, making UIST a very flexible tool for a variety of observations.

**Fig. 17.1** UIST under test by Suzie Ramsay in a laboratory at the UKATC in Edinburgh (Photo JAC)

Although the planned delivery date of April 2001 had by then slipped by about eight months due to the draining of staff to sort out problems with Michelle, an important milestone was reached in August 2001 when the UIST team obtained their first image in the laboratory. The target was a simple cardboard mask with the word UIST cut out, but it showed that the instrument worked and, although not the official UIST logo, this simple image was later used as the UIST icon on the JAC web pages since Chris Davis thought '*It was just so artistically and aesthetically pleasing*'. Unfortunately, the initial optimism following this success was short-lived since soon afterwards it was discovered that the array suffered from excessive read-noise. Despite determined efforts to reduce the noise, the array performance could not be brought to acceptable levels in time for the proposed February 2002 delivery. Resolving the noise problem, combined with the need for JAC staff to work on JCMT heavy engineering projects over the summer of 2002, forced a delay of several months to the delivery of the instrument. UIST, closely followed by its commissioning team of Suzie and her PhD student Stephen Todd, Mel, Jim Elliot (mechanical technician), Maureen Ellis (electronics engineer), Alistair Borrowman and David Gostick, eventually arrived in Hilo in July and August 2002.

UIST achieved first light on the telescope on the 24th of September 2002 signifying the end of an intense period of installation and testing by engineers and scientists from the UKATC and the JAC. Weather, telescope and instrument cooperated to allow UIST to be tested in its three major modes (imaging, spectroscopy and integral field spectroscopy) during its first half-night of observations. However, the satisfaction of seeing the instrument working on the telescope for the first time was shattered the next morning when the team woke up to find that calculations done while they were asleep had shown that the sensitivity of the instrument was lower than expected, with the real transmission being only 60 % of that predicted. The most likely culprits were either poor transmission through one or more of the optical elements or a lower than expected quantum efficiency of the detector.

**Fig. 17.2** The UIST commissioning team in the dome with UKIRT's new instrument. Front row, L-R: Chris Davis, Maureen Ellis, Suzie Ramsay, Alistair Borrowman. Back row L-R: Mel Strachan, Jimmy Elliot, David Gostick, Thor Wold, Russell Kackley, Mathew Rippa (Photo JAC)

Over the next few months about 20 nights were used for UIST commissioning, which was completed in December 2002 and shared risks observing began. Early observations covered the whole range of astronomy and included the planet Mars during its close approach to the earth in August 2003 and the quasar SDSS J1148 +525 which was at the time the most distant quasar known. The Mars observations were a big success. Chris remembers '*mucking about with some UIST images of Mars on a Friday afternoon* (*for a press release*) *and seeing them on the front of the London Times the following Monday*'.

One problem that did develop was with some of the grisms. These had been manufactured using a fairly innovative technology in that they were "replicated" rather than being ruled directly onto a KRS substrate in the traditional manner. In this process a layer of epoxy is spread onto the prism face and a master grating is then stamped onto it, leaving behind the required grating as an impression in the epoxy. While this had worked extremely well for some of the grisms, the shortest wavelength example had very poor transmission. In what Chris Davis described as '*A long and somewhat painful story*' another grism was ordered and installed in spring 2003. Sadly, this turned out to be little better, so a third example had to be ordered, this time based on more conventional technology. This at least turned out to be a little better than its predecessors although at these shorter wavelengths UIST never did perform as well as it did in the H and K bands.

The reduced throughput of UIST meant that for imaging in the near infrared, i.e. in the JHK bands, it was less sensitive than UFTI which remained in use while UIST was adopted for the other, more complicated modes of spectroscopy and thermal (L,M band) imaging which UFTI could not provide. In January 2007 a series of warm and cold tests of the optics were carried out which suggested that the transmission of the optics was indeed a little lower than expected, about 90 % of the prediction, but this drop was not enough to explain the overall sensitivity loss. So in the spring 2007 the detector was returned to Raytheon for testing of its performance. Raytheon were able to test the detector at $H$ and $K$, though their results were inconclusive and the problem was never fully resolved. Although other tests of the detector efficiency were considered, ultimately the switch to a dedicated WFCAM survey mode and the mothballing of UIST and the other Cassegrain instruments rendered the matter irrelevant in any practical sense.

UIST was gradually upgraded over the years, with various options for polarimetry being added. A second Wollaston prism was installed to make spectro-polarimetry possible with all of the available grisms. Also commissioned were modes for circular-imaging-polarimetry, circular-spectro-polarimetry, and capabilities for slitless spectroscopy. On the imaging side two coronographic slit masks were mounted into the slit wheel, making coronographic imaging or even coronographic-imaging-polarimetry possible. Chris Davis had the view that '*UIST's excellent imaging-acquisition software, developed early in the instrument's lifetime, meant that all of these exotic modes were ridiculously easy to operate, so much so that spectro-polarimetry was offered in service mode, and could be executed by almost any observer as part of flexible scheduling. All the modes were also supported by sophisticated pipeline reduction software*'.

There were also hardware improvements. The original EDICT array controller was not able to read out the entire UIST array quickly enough when working in the L and M bands, a wavelength region in which the sky is very bright. The array would simply saturate before all the pixels could be read out. Part of the solution was a new off-the-shelf array controller from SDSU, which was matched with a PC based acquisition system identical to those which controlled the four detectors in the Wide-Field Camera WFCAM. This provided both a significant improvement in the UIST readout and useful compatibility with the WFCAM systems. The other part was a modification to the way in which the array was read out. On-sky commissioning tests of the new readout method were successfully conducted in December 2007, finally making full-array read-out in the M' band possible.

UIST remained in use until all Cassegrain observing was stopped in January 2009 and it bowed out in much the same way as it had lived its life, as a versatile, multi-mode instrument with fabulous acquisition and data reduction software. On its last night a number of UIST programmes were flexibly observed and completed including long-slit spectroscopy, L-band spectro-polarimetry, L-band Integral-field spectroscopy, and a bit of straight-forward K-band imaging for good measure.

# Chapter 18
# The New Era of Wide-Field Astronomy at UKIRT

## WFCAM: The UKIRT Wide-Field Camera

By the mid-1990s the combination of the UK's decision to join the Gemini partnership plus a continuing financial squeeze meant that PPARC was looking for savings in its astronomy programme. Following on from the strategic review chaired by Professor Jim Hough in 1994, the Williams panel had noted that wide-field programmes at UKIRT were likely to become increasingly important, partly in direct support of Gemini, but also because this would be an extremely competitive use for an IR-optimised 4 m telescope in the era of 8 m and larger telescopes. The UKIRT staff were already aware of the threat from Gemini, and considerable thinking about the future had been going on for some time.

It was felt by many in both Hawaii and Edinburgh that the future of UKIRT lay in specialising the telescope for wide-field surveys and their follow-up. It was already clear that new generations of 2048 pixel square infrared arrays would soon be available and the existing infrared surveys, notably the 2MASS project, did not really go deep enough to provide worthwhile targets for spectroscopy with the new generation of 8–10 m telescopes. So there seemed to be considerable scientific potential for large area infrared surveys that could go much deeper than a J magnitude of about 15. So, although no-one can clearly identify the source of the idea, astronomers at the JAC and ROE began to set out the requirements for UKIRT to evolve into a wide-field survey telescope. An early manifestation of this was a paper written in 1997 to the astronomy committee by Suzie Ramsay, Tim Hawarden, and Mark Casali of the ROE together with Andy Lawrence and John Peacock from the University of Edinburgh's Institute for astronomy in which they set out key projects for a near infrared wide-field facility. They suggested modifications to UKIRT, notably a new secondary mirror, and new instrumentation for wide-field imaging and multi-object spectroscopy while retaining a capability for general purpose astronomy. The logic was simple, a wide-field UKIRT plus the UK share of time of the larger Gemini telescope would allow a coordinated approach,



J.K. Davies, *The Life Story of an Infrared Telescope*, Springer Praxis Books,
DOI 10.1007/978-3-319-23579-0_18

allocating projects onto the most suitable telescope for the scientific project in hand. The JAC management team in Hawaii were fully supportive of this proposal.

By the spring of 1998 the various UK committees (the Ground Based Facilities Committee and the Astronomy Committee) had agreed a plan for the UK's telescopes. UK participation in a large millimetre array, which would become ALMA, was the highest priority, with a wide-field UKIRT second. Various options were considered, including either splitting the infrared survey work between UKIRT and the Isaac Newton Telescope (INT; a 2.5 m optical telescope in La Palma) or taking the opposite approach and equipping UKIRT to provide optical as well as IR imaging. This latter option, while a considerable tribute to the quality of UKIRT's optics, would probably have doomed the INT and so was highly controversial. To help find the best outcome, there was an ad-hoc meeting in May 1998 under the chairmanship of Professor Mike Edmunds to discuss the future roles of UKIRT and the INT. This reported that the group could see no case for developing an optical capability on UKIRT and that there was a clear need for both wide-field survey science and independent science programmes in the infrared in the northern hemisphere. Eventually the dust settled, the capability for multi-object spectroscopy was dropped, and by 1999 a new UKIRT instrument was approved as a wide-field survey camera with the following characteristics:

*Four 2kx2k Hawaii-2 HgCdTe detectors laid out in a square. Four was the obvious number based on the likely cost and availability of the, yet to be built, detectors. One would not provide enough area to make a survey with reasonable speed and a three by three array of nine detectors would be far too expensive.*

*The detector size plus conditions of image quality and survey speed implied a field-of-view with a diameter of 0.93 degree, which was at the time very large by IR standards.*

*The survey would cover the z-K bands, ranging from the red end of the visible light window at 0.9 micron to just short of the wavelengths where increasing thermal background would start to become problematical.*

*The system would need a high bandwidth autoguider, operating at 20-100 Hz. This was required to continue to take advantage of the upgrades work and support UKIRT's secondary mirror tip-tilt system which had solved issues such as the windshake that had bedeviled UKIRT in its early years.*

*The existing Cassegrain instruments would need to be retained so any interesting discoveries not requiring the use of larger telescopes could be followed up using UKIRT itself.*

However there was one obvious, and very large, problem with this scheme. UKIRT had for years been operated as an f/36[1] telescope and as such its instruments typically had a field-of-view of a few arcminutes at most. Getting a field-of-view of almost a degree out of the optical system would be a very tall order. In fact, at first sight, it seemed impossible. For an f/36 beam to have the required field-of-view of 0.93° square, the focal plane would have to be 2.2 m in diameter and such a beam

[1] Although usually known as the f/35 focus, the original secondary mirror had been replaced during the upgrades programme and UKIRT actually operated with an f/36.32 focal ratio.

would not fit through the hole in the primary mirror to the support unit where UKIRT's instruments were traditionally mounted. Wide fields-of-view are often possible at a telescope's prime focus, but UKIRT had never operated instruments at this location for the entirely sensible reason that its prime focus was a point outside the small dome which had been specified 20 years earlier to save money.

The instrument would be built by the UK Astronomy Technology Centre at the ROE, where optical designer Eli Atad, with project scientist Mark Casali and others including optical engineer Dave Henry, went though many, increasingly crazy, concepts. They tried literally dozens of ideas until the team finally finished up with a design that involved a new secondary mirror and placed the putative camera on top of the primary mirror instead of underneath it. It looked completely ridiculous, especially at UKIRT with its well developed and highly flexible instrument support system mounted beneath the mirror.

This unusual location was driven by the sole requirement to have a wide field-of-view. Any kind of large optics would have to be made from fused silica and if these were lenses of any power they would introduce chromatic aberrations. Accordingly, all the power had to be in reflective optics, and so the design included an ellipsoidal M3 mirror. A new f/9 secondary mirror was designed to use the same attachment points as the existing f/36 secondary mirror. This meant that the new secondary could be mounted on the same piezo-electric tip-tilt stages for fast guiding. A lengthy study showed that the existing tip-tilt system would prove capable of handling this larger and heavier mirror with only a minor reduction in performance. Next in the optical path was a field lens positioned close to this mirror, then an ellipsoidal tertiary mirror which provided the optical power and further reduced the beam to f/2.4. However, all these optics introduced spherical aberrations so large that an aspheric corrector was needed to remove them.

The manufacturing of the optics proved challenging, but one by one the problems were overcome by either modifications to the design or by careful attention to detail when specifying materials and processes. The most troublesome component turned out to be the aspheric corrector, which almost proved to be a show-stopper when the manufacturer AMOS of Belgium found it hard to reach the required figure (shape) for the mirror. Eventually, and after a lot of pressure from Edinburgh, the final figure was reached, though well-behind schedule. In the end the corrector only caught up with the instrument in Hawaii a matter of weeks before the commissioning started.

The resulting design for the instrument was a tall, thin cone, looking not unlike the warhead of a missile and which (almost) fitted inside the shadow cone of the hole in the primary mirror. So, despite its size and location, the instrument would not actually block very much of the light from the sky. Putting WFCAM on top of the primary also had the nice feature of leaving the Cassegrain instruments (i.e. CGS4, UIST and UFTI) unaffected. So to switch to or from WFCAM it would "just" be necessary to change the secondary and put the instrument on or off.

The Conceptual Design Review for WFCAM was held in August 1999 at the ROE and Ian Robson attended from the JAC. Ian was really concerned at the concept of the camera being positioned above the mirror. He remembers confronting designer David Montgomery with the question '*so would you really stake your pension on the fact that when the telescope is pointing at an elevation of 25 degree, there is zero chance of the mounting ring, or the camera, breaking and as a result, smashing the primary mirror*?' David was very firm in his conviction and so Ian was content and the discussion then focused on the actual technicalities of the intriguing and novel design. The project was approved by PPARC in February 2000 with an estimated budget of £4.34 million. The completion date was set as September 2002. A preliminary design review took place on the 22nd and 23rd of November 2001 by which time delivery had already been pushed back to March 2003; it would eventually slip by two years. The cryostat was delivered to Edinburgh just before Christmas 2002 and vacuum tests were conducted at the UKATC in January 2003. The first cool-down, during which key internal temperatures were checked, took place in the spring of 2003. A major advance for the project was an agreement by which the Japanese astronomical community would gain access to UKIRT by paying for three of the four WFCAM detector arrays. This arrangement was agreed between Professors Ian Robson as Director JAC and Norio Kaifu, then the Director of the Subaru Telescope. However, although everything was positive, it took almost 18 months before the agreement was signed in June 2002 by Ian Robson and the new Subaru Director, Hiroshi Karoji. Under the deal, which saved PPARC over half a million pounds, Japanese astronomers would be assured access to some regular UKIRT observing until the surveys started and then would have opportunities to join some of the survey programmes once the new instrument arrived at UKIRT.

Development of such a large instrument presented a number of engineering challenges, not the least being how to fit the detectors and the telescope's autoguiding system into the available space. The HgCdTe detector arrays chosen were not buttable so they had to be physically separated to allow space for their connectors and wiring. This meant a much larger focal plane, which in turn needed more difficult and larger optics, resulting in a larger and more expensive instrument. It would also require a complex observing mosaic pattern to cover an area of sky without gaps. Two-edge buttable arrays were under development and just about available, but the timescale made it sensible to choose the existing arrays rather than wait for something better. Indeed, there was considerable pressure from some quarters to select an alternative array design, but the team correctly decided that even taking into account delays to the WFCAM project, the timescale for the materialisation of any better detectors was also most probably optimistic. In the event the decision to use the existing arrays was vindicated. A wide spacing of the four detectors was chosen, with the autoguider CCD at the centre of the pattern but lifted slightly above the IR detectors and rotated by 45° to make the most use of the available space. This put the autoguider out of focus, so an extra central field lens was used to raise the focus of the light falling onto the CCD so that it and the infrared detectors would all be in focus at the same time.

Alignment of the instrument on the optical axis of the telescope would be crucial to getting the desired optical quality, but the only place strong enough to mount this huge instrument was the central lifting ring of the primary mirror. The problem was that no-one knew how accurately this ring, which surrounded the central hole in the primary, was centred on the actual optical axis of the mirror. Luckily an old Grubb Parsons report was found which indicated that the centring was good to about 4 mm.

Gradually the system started to come together. However, as with many projects, the main problems turned out to be completely unexpected. Probably the most serious was when one of the IR detectors disintegrated during instrument testing. Infrared detectors use hybrid materials and are designed to deal with the large internal stresses which result from differential contraction when they are cooled to their very low operating temperatures. However, this was something that was thought to be well understood, so it was a great shock when two different detectors violently delaminated in the test cryostat. This was a major setback and the project was delayed for six months as the WFCAM team tested and studied their thermal design and cooling strategy.

Getting more information on the problem was not simple; the detectors were too expensive to be wantonly tested to destruction and the manufacturer, Rockwell, was naturally anxious that the issue not be discussed too openly within the astronomical community for fear of damage to its reputation. According to Mark Casali '*Rockwell asked us to choose our words carefully, and especially not to use the word "exploded" in describing the detector failures. Since our relationship with them was crucial for both WFCAM and future instruments, we subsequently used the less graphic word delamination in public presentations and documents*'. The team went to some lengths to determine what they thought was happening, even inspecting the devices under high powered microscopes and making drawings of what they believed to be the details of the array structure to identify where the problem might be occurring. This was helpful in talking with Rockwell and finally, after lengthy discussions and a long series of tests, it was concluded that the problem was due to contamination of the detectors during manufacture. The detectors were replaced and fortunately no further problems of this nature were experienced.

## UKIDSS

While the engineering teams in Edinburgh and Hawaii were wrestling with the problems of what kind of wide-field facility UKIRT would have, seeking funding and then actually designing and building this huge new instrument, another group of astronomers had been busy trying to decide how best to use it. Andy Lawrence recalls that when what became WFCAM was first mooted, it was seen as a wide-angle facility for the community to do many medium-sized surveys, rather than to do a single giant one. However the traditional free for all of telescope proposals and six monthly reviews was unlikely to be efficient nor to maximize the scientific output from the huge field of view offered by WFCAM. Recognising this he, together with Andy Adamson, Tim Hawarden, John Peacock, Will Saunders, Nigel Hambly and Mark Casali wrote a proposal for a single large, multipurpose survey. This, which was already close to the concept that would eventually be approved, would require about 50 % of the observing time at UKIRT for between five and seven years. The scheme, as an unsolicited paper for information entitled *A PROPOSED LARGE PUBLIC SURVEY WITH THE UKIRT-WIDE-FIELD FACILITY* was put to the UK's Ground Based Facilities Committee, an advisory body covering the range of the UK facilities, in September 1998.

**Fig. 18.1** The WFCAM cryostat in the ROE south building laboratory (Photo UKATC)

The response was encouraging, but setting aside half a telescope for a good fraction of a decade to a single project was a big decision and so the proposal, now called the UKIRT Infrared Deep Sky Survey (UKIDSS) and with an expanded list of participants described as a proto-consortium, was put to the UKIRT management board in November 1999 to seek their approval in principle. The plan envisaged the group of proposers running the whole survey on behalf of the community, keeping them informed through announcements and being overseen by a science advisory group of some kind. Various details of management, survey duration, data rights and co-operation with the Japanese astronomical community who were expressing an interest in WFCAM were set out. Finally, the board was invited to invent a better acronym than "UKIDSS".

The response was in effect to say "We like the idea, but it has to be opened out to the rest of the UK". Accordingly in July 2000 there was a large meeting in Leicester from which evolved a consortium to design and undertake a set of complementary surveys. The consortium included about 60 UK astronomers plus a few Japanese scientists who were invited to join in order to promote scientific collaboration between Japan and the UK in the hope of joint exploitation of UKIRT and the Japanese 8 m telescope Subaru, by now fully operational. This consortium proposed a wedding cake of projects, each going deeper but over a smaller area than the layer below and eventually comprised a set of five complementary surveys ranging in depth and area from K = 18.4 over 4000 sq.deg to K = 23 over 0.77 sq.deg. This would be much deeper than the 2MASS survey, and the volume surveyed more than an order of magnitude greater. The entire programme was expected to take about seven years to complete. The consortium appointed Steve Warren as UKIDSS Survey Scientist and the UKIRT board approved the basic survey idea combined with a rolling review process to monitor progress. The five nested surveys were:

***A Large Area Survey* (*4000 sq.deg*, *K* = *18.4*)** The LAS, covering high Galactic latitudes, would provide a good match to the Sloan sky survey in the visible wavelength region and the combined database would provide accurate photometry of galaxies in the local universe over the wavelength range 0.35–2.3 μm. However the LAS was not just about galaxies, it would explore the entire observable Universe from nearby cool stars out to the most distant quasars at redshifts as high as z = 7.

***A Galactic Plane Survey* (*1800 sq.deg*, *K* = *19.0*)** The GPS was intended to map a strip 10° wide covering half the length of the Galactic plane. The survey would be built up in three observing cycles, providing information on variability of all the sources observed. The survey would enable a proper census of the contents of the galactic disk and provide a catalogue for the identification of sources detected at other wavelengths, for example X-ray binaries detected from space. Since most star formation occurs in the galactic plane, the survey had the potential to increase the number of known Young Stellar Objects many fold.

***A Galactic Clusters Survey* (*1600 sq.deg*, *K* = *18.7*)** The GCS would study the sub-stellar mass function by targeting large open star clusters and star formation associations. By discovering large numbers of low mass and Brown Dwarf stars the GCS would provide an accurate measure of the Initial Mass Function below the hydrogen burning limit for a range of star clusters with different ages and metallicities. This would allow astronomers to determine if, and how, the star formation process varied from cluster to cluster.

***A Deep Extragalactic Survey* (*35 sq.deg*, *K* = *21.0*)** The DXS was designed to compare the properties of the Universe in the past with its properties today. It was intended to measure the abundance of rich galaxy clusters at redshifts between 1 and 1.5 and to measure galaxy clustering at redshifts greater than 1. The fields chosen for the DXS would be targets of intensive multi-wavelength campaigns by other groups.

***An Ultra Deep Survey* (*0.77 sq.deg*, *K* = *23.0*)** The UDS was intended to map 0.77sq.deg of sky to a depth of K = 23 by repeating a single pointing many times over and combining all the images. This would provide the first detailed picture of large scale structure at a redshift of 3. The combination of depth and area had the potential to make this the most important existing archive of near-IR data for statistical studies of the early stages of galaxy formation.

The final proposal was made in March 2001 and, after some iteration on the details of the individual mini-surveys, was accepted by the end of the year. The UKIRT board never did invent a better name, and UKIDSS it remained.

## Other Preparations

As the instrument began to come together at ROE in 2003 other groups were busy preparing for the day that WFCAM would be installed on the telescope. The Astronomical Survey Unit at the Institute of Astronomy in Cambridge was busy developing a data reduction pipeline, which would run in the UK, and were also building a simpler one to run at the telescope for near real time data quality assurance. The Wide-Field Astronomy Unit at the Institute for Astronomy in Edinburgh were hard at work developing the science archive and user interface tools to query the large catalogues and image data that would be produced. The JAC in Hilo took on telescope software developments for WFCAM, the design of the WFCAM interface plug in the primary mirror and the design and procurement of the WFCAM handling system.

## WFCAM Commissioning

The size and mass of WFCAM presented a whole new raft of challenges for the staff in Hawaii who would have to install the instrument on the telescope, not once, but regularly. This was because the basic plan for the use of WFCAM was to operate it for long blocks of time, typically six to eight months, and then remove it so that the existing instruments mounted below the mirror could be brought into action for a few months of follow-up and other traditional one-off projects. So a method needed to be developed that was reliable, safe and could be done at least a couple of times a year without losing too much observing time. This was not a trivial problem, WFCAM weighed more than a tonne and had to be lifted above the primary mirror and then lowered with considerable precision onto the central mounting point. Aside from the risk of injury to the staff who would be doing the job, dropping the instrument onto the mirror would most likely be catastrophic for both WFCAM and UKIRT. The worst case scenario would be an earthquake occurring at a critical point during the lifting process and mitigating this risk meant that the whole process had to be engineered with a large margin of safety against this, albeit very unlikely, event.

According to Mark Casali '*Early concepts included assembling jib cranes on the telescope or using a system of winches to lift it into place. In the end all of these ideas were rejected by the JAC engineers* [*notably Chief Engineer Dean Schutt*] *who took the lead in solving the problem and found that a specially modified fork lift truck from Hyster would meet our requirements. That the solution to our problem lay in asking one of the World's largest lifting-equipment manufacturers to solve it for us should perhaps have been obvious earlier*!' Testing of the modified fork-lift was done using a specially made full size polystyrene model of the WFCAM cryostat. Initial tests took place in Hilo using a spare mirror plug supported at the same height above the ground as the real mirror plug is above the dome floor at UKIRT. Tests were also done using a metal block that weighed the same as WFCAM. Following success in Hilo, testing moved to the UKIRT dome where the fact that the dome was barely bigger than the telescope and that the gaps between the telescope struts are only marginally larger than the WFCAM instrument itself made it clear that manoeuvring the truck into position and lifting the instrument up through the telescope struts to put the instrument precisely into its mounting would be a very delicate procedure. Nonetheless, the test showed that it could be done and another milestone was passed.

WFCAM was shipped from Edinburgh and all 14 of the crates safely arrived at the JAC on the 17th of August 2004. The WFCAM team arrived a week or so later and the unpacking and assembly of the instrument in began Hilo. The instrument had been shipped with all the optics and the detectors packed separately for safe transport, so the first task was to reinstall them at the JAC. After this was done WFCAM was transported to UKIRT on the 14th of September, and mounted onto the Michelle manipulator in the dome the next day. Warm functional tests were carried out the following week and these showed that all was in order. A few more weeks were required for the installation of some delayed components, cold testing and finalisation of the installation procedure before WFCAM was finally mounted on the telescope.

**Fig. 18.2** Testing the mounting procedure using the polystyrene WFCAM model (Photo JAC)

The first measurements with an alignment telescope showed a problem; the secondary mirror was badly misaligned. This was very puzzling, because WFCAM had been designed and built with a great deal of attention to alignment error budgets and component specifications. Examination of the secondary mirror with a pair of binoculars revealed the terrible truth. The mirror support fixtures, glued on by the manufacturer, had broken off and the only thing stopping the mirror from crashing onto the primary mirror (and heads) below was an emergency support ring, designed in for safety. While the team pondered sending the mirror back to SESO in France for re-gluing, Tim Chuter from the Joint Astronomy Centre purchased cleaning materials and a specialised epoxy glue and proposed that he would glue the mirror back in place himself! This was eventually agreed by the team and the result was very successful, the mirror is still securely glued in place today.

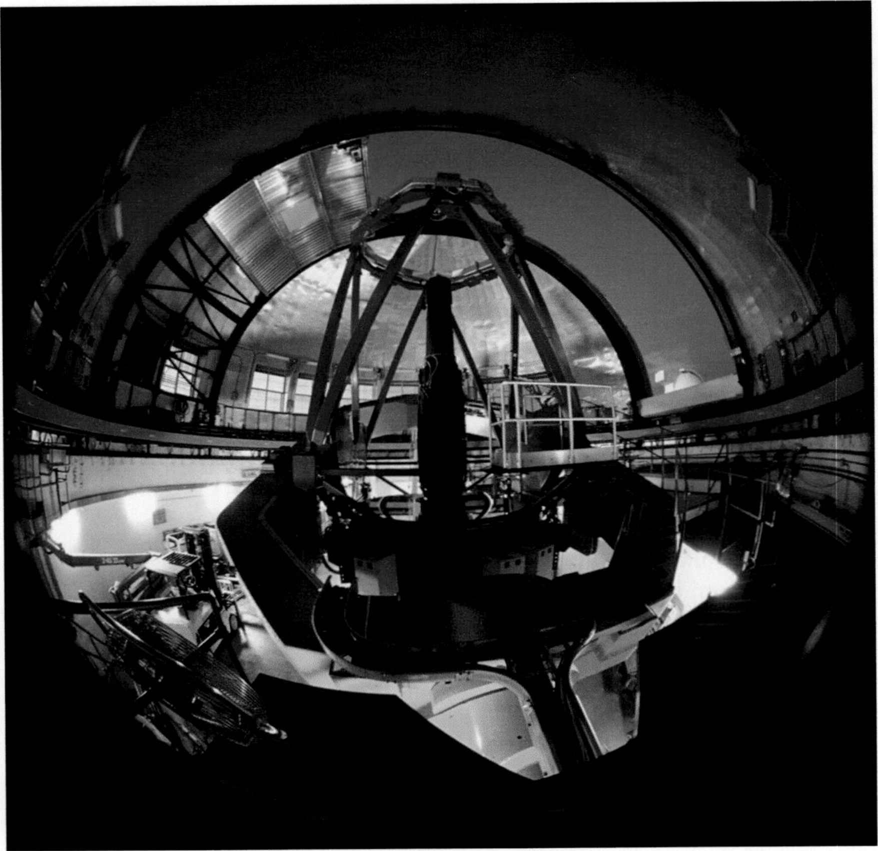

**Fig. 18.3** This fisheye lens shot shows WFCAM mounted on UKIRT (Photo Paul Hirst)

First light with the instrument was achieved on the 21st of October 2004. According to Mark Casali '*The first image was tremendously reassuring*'. The stars were approximately 1 arcsecond in size, and were round and relatively uniform over the entire field. This indicated the optics were well aligned with no major residual astigmatism. Mark's presence was fortunate because he had recently been offered a new position at the European Southern Observatory headquarters in Garching as Head of the Infrared Instrumentation Department. In principle he was due to have left the UKATC before the commissioning started, but ESO agreed to him taking up his new job after the WFCAM commissioning. He came back to Europe in September 2004 and started his new job at ESO in October. By then the local JAC team had taken over the commissioning with the UKATC lending support as required.

Of course the first night of observing was only the beginning of the commissioning process and there was much work to be done before WFCAM could start its real survey work. Much progress was made over the next few weeks, but the first commissioning run was constrained by adjustments that were impossible to fit in before the resumption of normal observing with the regular instruments. Nonetheless a superb image of the Orion region obtained during the commissioning indicated the potential of WFCAM very early on. Mark Casali points out that '*Over the difficult years of design and assembly perhaps the most remarkable thing is that no one (including our funding agency PPARC) ever said the design was too ambitious and would never work. But looking back now I realise we may have included too many new developments in one instrument, and the project faced major difficulties on several occasions*'.

WFCAM was reinstalled on the telescope in March 2005 for further commissioning work. This showed up issues with the alignment of some components, which were temporarily resolved at some sacrifice to image quality pending a final resolution on the next warm-up in June. Full survey operations began on the 13th of May 2005, and the data taken in the first observing block up to the end of June provided about 2 % of the planned seven-year survey. Particular fields were targeted for science verification purposes and these data were pipelined and archived before being analysed and the results fed back to the groups operating and developing the data reduction software. After a further set of internal adjustments the instrument returned to the telescope in August and survey operations were soon underway.

At first the data taken was sent back to the UK on LT01 data tapes via Federal Express. The tapes were written in sets of four, writing all four simultaneously, one per WFCAM detector, because the computers and tape systems were not fast enough to guarantee that all the data from one night could be written to a single tape before the next night's observing started. Paul Hirst used to pack the tapes in sets of four in jiffy bags, double bagging them for luck, and writing "scientific data on magnetic tape" in big letters on the bag, hoping that this would avoid them being x-rayed too many times by overzealous security officers and would also avoid customs charges. If the weather was good and WFCAM was in constant use, the JAC would be sending out a set of four tapes about once a week. As much as possible was automated, if the tapes filled up they would automatically eject and an email would be sent to the staff at the summit to put in a new set of tapes and how to label them. Later, purely electronic transmission direct to the data processing team at Cambridge replaced the tape-based system.

The co-operation with the Japanese astronomers, although cordial, was not always straightforward to support because of language and cultural difficulties. According to one staff member a typical Japanese run took about the same amount of effort to support as two or three regular UK runs. Another scientist feels that, while language was sometimes a problem and that '*at times the TSS and the observers were communicating by cartoons*', the large teams that the Japanese tended to bring to the telescope was perhaps a bigger issue, often sending four observers when one would have been sufficient.

Access to WFCAM data had been promised to the ESO community as part of the conditions for the UK's accession to the organisation and the first ESO-wide release of UKIDSS data took place on the 10th of February 2006. This was really just a small prototype dataset and the first large release of UKIDSS survey-quality data was made on the 27th of July the same year. In the latter part of 2007 WFCAM reached a milestone with the observation of its one millionth frame. This particular frame was not part of the UKIDSS programme, but was taken for one of several other projects that ran in parallel with the survey work. It was part of a project to search for Earth-sized planets and was one of four large campaigns which had recently been approved as part of a re-alignment of UKIRT operations away from traditional small projects and towards a smaller number of larger programmes. Brad Cavanagh from the JAC software group, wrote that '*A million WFCAM observations certainly does add up to a lot of data; a hair over 61 TB of data, to be exact. If you wanted to put all of the data on CDs you'd need to buy 98,461 of them, and they'd make a stack over 387 ft (118 m) high*'.

For the first few years of WFCAM operations, UKIRT operated several months of survey operations before stopping and switching back to the Cassegrain instruments for a few months of more traditional observing. For each switch it was necessary to remove WFCAM and then turn on the other instruments and bring them back into service. Overall this went well, taking about 3 days with the telescope closed for mechanical work followed by 2 nights of engineering time setting up and re-aligning the instruments. Overall it all went very smoothly. Unsurprisingly the TSS team found that they tended to be rusty for a night, but things came back to them very quickly.

According to Tom Kerr in 2012, eight years after it had been first installed, WFCAM had proved to be very reliable. The main problem was that the filter system is very complicated and the filter paddle mechanisms had begun to show signs of wear and tear. Since by that time the JAC did not have the effort to repair the mechanisms properly, they switched the most common and least commonly used filters around, putting the most popular filter into the paddle that had had the least wear over the years. There were also a few failures of the compressors in the coolers and to reduce the risk of these the UKIRT staff adopted a schedule of annual WFCAM maintenance, closing the telescope for two or three weeks and overhauling the coolers, changing filters and so forth.

It was during one of these annual downtimes that Tom Kerr had his own "14,000 ft moment". UKIRT agreed to host a visitor instrument called CELESTE, a mid-Infrared spectrometer. So they put back the old Cassegrain secondary mirror and set up CELESTE. Tom says it was '*Great fun and good to be back at the summit again*'. However on one of the last days the team drove up to the summit do a cryogen refill using the CELESTE team's own hire car and, when they reached the dome, Tom realised that he did not have a front door key. A door key is on the key rings of all the JAC cars so it is not something that is usually necessary to think about but, since this was hire car, there was no dome key on the ring. So they just had to drive down to Hale Pohaku to pick-up a key then drive back up again.

## UKIDSS Operations

Before leaving WFCAM it is worth noting that the survey operations were carried out by a combination of visiting observers from the UKIDSS consortium and JAC support scientists who in practice took about 30 % of the data. This placed new demands on the local staff. They still had their traditional role of welcoming visiting observers, bringing them up to speed on the instrument and observing procedures, but there was also the job of monitoring progress and setting priorities to be done. According to issue 23 of the UKIRT newsletter '*Each UKIDSS sub-survey has been split up into smaller projects, each dealing with either a particular area of the sky, or relating to a specific semester or observing period. The number of these projects has now exceeded 130! Each project has particular observing conditions that need to be met, and many of the projects are linked to others, meaning that they benefit from being observed in a specific order. The automatic observations scheduler takes care of sorting out the problems related to observability and sky conditions constraints. Still, this is not enough to ensure that the survey is carried out according to the original project, and in the most effective way. During weekly meetings, the overall status of the survey for the current semester is analyzed, and priorities are assigned to individual projects to ensure that they get done in the order in which the survey heads have planned. Potential conflicts between different projects are common, and they need to be resolved by either postponing lower priority projects, or by making sure that an equal amount of time is spent on each of them*'.

Andy Longmore had experience of the whole process from a number of perspectives, as a member of one of the survey consortia, as a visiting observer and as a manager who had access to the weekly reports. In his opinion '*The level of dedication, broad awareness of the programme objectives and hard work needed to ensure that this prioritisation process worked effectively was immense*'. According to Andy '*It is a huge credit to the UKIRT support staff, and particularly Andy Adamson and Tom Kerr, that the observer always seemed to have good guidelines with which to work*'. However, as UKIDSS worked on, passing the milestone of 100 published papers in August 2009, storm clouds were again gathering over UKIRT.

# Chapter 19
# Into a Third Decade

## Exodus

The period around the arrival, commissioning and early operations of WFCAM marked a period of considerable change amongst the staff in Hawaii. First to leave was Dean Shutt, who had been the JAC chief engineer for four years and who retired in July 2004. Simon Craig, who had held posts at both UK Royal Observatories and had also spent several years in Hawaii in the 1990s, was recruited to take his place. Also on the engineering side, David Laird returned to Edinburgh after a three year tour of duty as an electronics engineer. Chris Yamasaki and Kevin O'Connell did not have so far to go, they moved across the road to jobs at Gemini and the Smithsonian Millimetre Array, respectively. Desiree Milar-Okinaka, whose hula dancing had graced a number of JAC events over the years, left the administration department of the JAC to start a children's day care centre.

The scientists were also on the move. Marc Seigar left in September 2004 for California and the same winter Jane Buckle took a postdoctoral position in Cambridge. Jane was replaced as UKIRT telescope scheduler by Mark Rawlings. Olga Kuhn, who had spent almost five years at the JAC in one of the split PDRA/TSS positions left to become a support scientist at the Large Binocular Telescope in Arizona, her seat at the summit being filled by Lucas Fuhrman from the University of Northern Arizona. Lucas would not stay for long, after two years he left for Gemini South in June 2007, and was replaced by Jack Ehle in August. Another major loss was Nick Rees, the hardworking Head of software at the JAC, who moved to the UK to work at the Diamond Light Source in Oxford.



J.K. Davies, *The Life Story of an Infrared Telescope*, Springer Praxis Books,
DOI 10.1007/978-3-319-23579-0_19

## 25th Birthday

Late 2004 marked the 25th anniversary of both the dedication of UKIRT and the almost simultaneous opening of the nearby NASA IRTF. To celebrate a quarter-century of infrared astronomy on Mauna Kea a joint event was held in Hilo, both to share the organisational load and to celebrate the links between the two observatories. The main event was an evening drinks reception followed by a buffet dinner with speeches. A lot of good food was had, and many old acquaintances were renewed. Richard Ellis gave a very welcome and positive speech on the history of UKIRT and talked a little about its future, Eric Becklin did the same for the IRTF. John Jefferies, long since retired from the University of Hawaii, reflected on the differences between the pioneering atmosphere of the 1970s and the more regulated environment of the twenty-first century. The following day, the University of Hawaii held a session with two talks followed by a panel discussion at which Terry Lee gave an historical overview of the planning for and construction of UKIRT. For the first time ever UKIRT held an open day at the summit. While more than 100 people came to see the telescope and to hear about the challenges of doing infrared astronomy there was a moment of minor drama when a member of the public collapsed inside the dome during the tour. After suitable first aid, the visitor was taken down the mountain to meet an ambulance.

## Deep Impact

On the 4th of July 2005 the NASA *Deep Impact* mission aimed a dustbin-sized block of copper at the nucleus of comet Tempel-1. The objective was to simulate a meteorite impact on the comet and eject material from below the surface to form a crater. For a couple of anxious minutes around 7:52 pm Hawaii Standard Time NASA's Jet Propulsion Laboratory in Pasadena waited with bated breath for news that they had struck home. The first to get them that news was UKIRT where Steve Miller, Tom Stallard and Bob Barber, from University College London were being supported by Paul Hirst and Tim Carroll. Just a few seconds after the predicted time the telescope's fast guider camera detected a flash as the impactor struck home. The team immediately got onto the telephone link with NASA and reported that impact had occurred with the comet doubling in brightness in the course of just over a minute. Ironically, one person in the team did not get the news quite so quickly. Software engineer Maren Purves was sitting in the Hilo office on standby in case of computer problems and missed out for a while. The comet continued to brighten steadily for the next hour as gas and dust were blown out from the impact site to form a large cloud on one side of the nucleus. The whole process was also observed by the cameras onboard the main Deep Impact spacecraft, as well as by telescopes around the world. UKIRT also took several CGS4 spectra of the gases blown out from the comet after the impact.

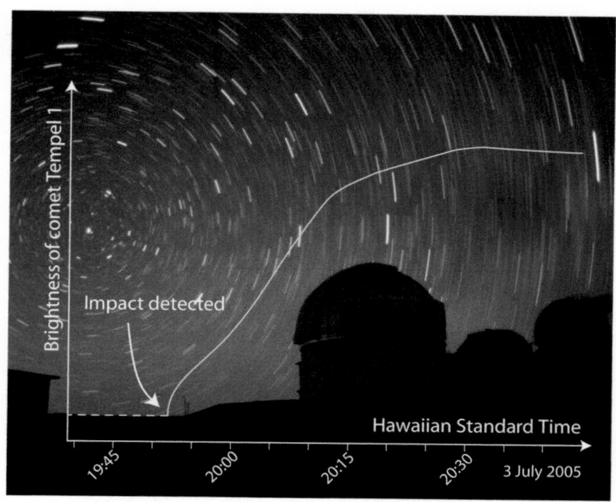


**Fig. 19.1** The brightening of comet Tempel-1 after the Deep Impact event as captured by the UKIRT guide camera. The data from the camera is superimposed on a time exposure of the UKIRT building. The constellation of the Southern Cross is visible on the *left* (Photo JAC via Douglas Pierce-Price, background image courtesy of Nik Szymanek)

## A Bad Winter

The winter of 2005–2006 was notable for two, rather different reasons. For one thing the six month long observing semester 2005B was the first in the history of UKIRT that was totally devoted to just one instrument as efforts were made to push forward with the UKIDSS surveys, This created an interesting problem for the TSS team, who would go for many months without using any of the other instruments and then, when the WFCAM was removed for the next sequence of Cassegrain observing, had to remember how to run UFTI, UIST and CGS4 in their many and complicated operating modes. Thor Wold wrote that '*We three TSSs had enough fits trying to once again remember how to run the three Cassegrain instruments after the 13 weeks of this last WFCAM shakedown period. It shall be most interesting when we have to encounter this once again - but this time after a continuous period of six months*!'

It was also a very bad winter, 2006 delivered the wettest March in Hawaii for 55 years and the mountain suffered weeks of winter storms as a series of upper-level lows and associated fronts marched across the Pacific and brought fog, ice and snow to the summit. At times the snow drifted up to six feet (2 m) in places between Hale Pohaku and the summit ridge and Thor Wold lost four straight shifts, each of five nights, without ever reaching the summit. For weeks, except for the occasional foray to the telescope to see how bad things were, the staff who were on duty were effectively imprisoned at Hale Pohaku.

## Return of the Sky at Night

January and February 2006 brought TV crews back to UKIRT. A film crew from the BBC Sky at Night programme arrived a few weeks after Christmas 2005 and spent a very busy week filming everything they (and the local staff) could think of. Thor

Wold was the on duty Telescope Specialist and spent considerable time moving the dome and adjusting the telescope to get just the right shot that the TV people required. A number of UKIRT staff members were interviewed and spectacular day-time and night-time views were taken of the telescopes, the scenery, dawn on Mauna Kea and the moon and stars shining down on the telescope domes. A wonderful evening sequence was captured showing the UKIRT dome opening and rotating while the sun set in the background. The programme was shown on the BBC in early February. As a summary of life on Hawaii's tallest mountain, the exciting science being done, and the present and future instrumentation being developed, the show was described by Andy Adamson as '*A great success*'.

No sooner had the BBC team left then another team arrived, this time for a series of astronomy events at local schools. The "*Journey through the Universe*" was an educational initiative led by the Carl Sagan Center for Earth and Space Science Education and involved many of the telescopes on Mauna Kea. Ironically in view of this pulse of publicity, at about the same time Douglas Pierce-Price, who had held the public outreach post since late 2001, moved to a new position as Education Officer at ESO and was replaced in March 2006 by Inge Heyer from the Space Telescope Science Institute in Baltimore.

In the meantime, the exodus of experienced UKIRT staff continued. Neal Masuda, a stalwart of the engineering division since 1990, and Chris Yamasaki left to work at Gemini. Mathew Rippa, after years of wrestling with the UKIRT Observatory Control System, switched to developing software at Gemini in March 2006. Two support scientists, Sandy Leggett and Paul Hirst, joined him at Gemini in late 2006. Sandy had been at JAC for more than a decade working on IRCAM, the thermal upgrade to TUFTI, the commissioning of UFTI and then the arrival of UIST. Paul had been at UKIRT since 2000 and was largely responsible for the spectroscopy part of the ORAC-DR pipeline. He was also central to the commissioning of the WFCAM and for UKIRT's relatively smooth transition to survey observing. Paul remembers that he '*Moved to Gemini in September 2006. When I got my tour of the Gemini dome during the day as part of the interview, we walked into the dome and there was Chris Yamasaki and Neal Masuda up a ladder under the telescope working on an instrument. Made it feel like maybe it could be a home from home...*'. Another departure for the UK was one of the remaining "old lags" Ian Midson, who had been in the administrative staff in Hawaii for 15 years, mostly acting as human resources manager and who departed in August 2007. Ian's almost infinite resources of patience and good humour had smoothed many a transition from the UK, and the US mainland, into the very different culture of the Big Island.

The holes in the support scientist team were filled by Watson Varricatt, returning to UKIRT where he had been the other member of the second generation PDRA/TSS and Luca Rizzi. Watson became responsible for (among other things) imaging with UIST and UFTI, while Luca took over as Sky Survey Support Scientist, managing UKIDSS from the JAC side of the world. About the same time Sam Hart's arrival bolstered the software group.

# Chapter 20
# UKIRT Under Threat

## Strategic Review

In 2005 the UKIRT board had decided to carry out a strategic review of UKIRT to '*Consider the current scientific utility of the telescope in the context of other developing infrared facilities worldwide and to evaluate the prospects for its continued exploitation over the next 10–15 years*'. The sub-text of this review was that while UKIRT had been an admirable front-line general purpose telescope for much of its life, the availability of infrared instruments on bigger telescopes, notably those of the ESO VLT and Gemini, and the construction of dedicated infrared survey telescopes such as VISTA, called its long-term future into question. The review was carried out by an independent panel of six international experts chaired by Professor Richard Ellis. The process began with a community discussion at the 2005 National Astronomy Meeting at which a wide range of possible futures were put forward. In June, the panel visited Hawaii and toured the Hilo and summit facilities before preparing their report in July and August.

The review was very positive about the telescope and its operations saying "*The panel considers UKIRT to be a remarkably efficient and well run observatory. Our panel is strongly of the opinion that UKIRT has a vital and promising scientific future as a 4 m telescope on a strategically important Northern hemisphere site*". Ellis and his panel also gave a nod to the history of the telescope and its ability to adapt to changing circumstances adding "*The UKIRT story is a truly remarkable one. Initially conceived as a dedicated infrared "light bucket" . . . via WFCAM and its associated optics, UKIRT represents the current frontier.*" The report also considered the future beyond the era when the UKIDSS work was complete, remarking on the possibility of upgrading the WFCAM with better arrays, the potential for installing an adaptive optics system and new instrument concepts such as wide-field spectroscopy and searches for extra-solar planets.

The Ellis panel's main conclusion was that there were several exciting niches for UKIRT in the future, although none of these ideas had yet been developed to the



J.K. Davies, *The Life Story of an Infrared Telescope*, Springer Praxis Books,
DOI 10.1007/978-3-319-23579-0_20

point where they could be rigorously evaluated. As noted earlier, and recognising the need to make economies, the report recommended that UKIRT concentrate on campaign science and proposed to the UKIRT Board that PPARC should invite proposals for ambitious three to five year observing campaigns requiring several hundred nights of UKIRT time to address fundamental high impact science questions. These campaigns could involve WFCAM or new, dedicated, instruments. There turned out to be a remarkable demand for such campaign time. Twenty proposals were received, five of which were for renewal of the UKIDSS surveys which the Board wanted to include in the competitive process to ensure that UKIRT continued to do front line science. The total amount of time requested by all these proposals would have kept the telescope busy for nine years!

Despite the positive outcome of Ellis's review, and the demonstrably high community demand for UKIRT, in May 2006 PPARC decided to reduce its financial commitment to UKIRT operations to roughly 50 % of its current level (i.e. from about £2 million to under £1 million per year) by 2010 at the latest. Failure to achieve these savings would lead to closure. A budget reduction of this magnitude was well beyond the scope of internal savings, and it was obvious that both a reduction in staff numbers and a new partnership would be required if UKIRT was to have a future beyond 2010.

Japanese astronomers had been buying a few UKIRT nights per year since 2005, continuing the arrangement which had begun when they provided several of the WFCAM arrays in return for telescope time, and some money was coming in from the European Commission, but it was not enough to make a huge difference. Accordingly, Gary Davis immediately increased his efforts to find new partners, looking both towards the east (India, China, Japan, Korea) and within the USA. UKIRT hosted a delegation from China in 2007 and Gary had lengthy discussions with astronomers from India and the California Institute of Technology, but none of these came to anything. The talks with the Koreans were, however, eventually to bear fruit. In 2008 the Centre for the Exploration of the Origin of the Universe (CEOU) at Seoul National University agreed to purchase a modest amount of telescope time over the coming three years and a memorandum of understanding was signed during the UKIRT board meeting in December 2008. However, before this could happen, things were about to get a whole lot worse.

In 2007 PPARC was merged with the CCLRC[1] (from which it had been cleaved a decade or so earlier during an earlier research council restructuring) to form the Science and Technology Facilities Council (STFC). This merger was trumpeted as an opportunity to improve the situation regarding research opportunities in astronomy and particle physics, but the result was, by most reckonings, a disaster for UK astronomy. One of the first actions of the newly formed STFC was to establish a programmatic review to see how the programmes of the two councils could be fitted

[1] CCLRC was the rather artificially named "Council for the Central Laboratory of the Research Councils", basically the Rutherford and Appleton Laboratory near Oxford. It had been part of SERC before the demerger of the research councils in which PPARC was created.

together. While this process was underway the outcome of the UK government's 2007 comprehensive spending review settlement (which set the budget for the next few years) was announced. This settlement provided what Peter Warry, chairman of the newly formed STFC, called '*The most difficult outcome for the Research Councils since 2000 and would inevitably have serious implications for STFC's programme over the next four years*'. The practical implications were spelled out by Professor John Womersley, the STFC's Head of programmes, on the 22nd of November 2007 when he said '*While the settlement will enable the Council to pursue much of our planned programme, the costs of running the STFC will increase not just with inflation but also due to the increased costs of operating some new major facilities. The consequence is that with other minor adjustments the STFC is looking at a deficit of about £80 million in its existing programme over the CSR period*'.

Thus the programmatic review took on greater urgency as STFC attempted to find how it could save this enormous sum of money. Prioritising the astronomy programme fell into the remit of one of the recently formed STFC panels which covered the fields of Particle Physics, Astronomy and Nuclear Physics (PPAN). There were only two astronomers on PPAN and, when the first version of its conclusions were published in March 2008, it produced an astonishing result for the UK infrared astronomy community, UKIRT was ranked in lower priority category, the bottom of the four priority bins. The baseline plan from the programmatic review was to move UKIRT to 100 % survey mode from the 1st of April 2008 in an attempt to complete as much of the survey work as possible while efforts continued to find international partners to share the cost of running the telescope. Should new partners fail to materialise, immediate closure would be seriously considered.

The outcry over the result of the programmatic review from across the physics and astronomy community forced STFC into a rethink. Although critics of the proposed prioritisation exercise came from across the board, the response from the UKIRT community was described politely as being "*High in numbers and robust in tone*", although aghast and outraged might have been an equally good description. As a consequence, yet another round of reviews, this time by expert panels with wide community consultation, was ordered and it was decided that UKIRT could remain operating as a general purpose facility until the end of 2008, when the 100 % survey option would be revisited. Professor Martin Ward of Durham University chaired the panel looking at ground-based telescopes and in the autumn of 2008 recommended that '*Given the high priority of UKIDSS, recognized by both the panel and strongly emphasized by the community inputs, the panel recommends raising the priority of UKIRT from band 4 up to band 2*'.

In the meantime, the operations team in Hawaii tried to make the best of things in an increasingly gloomy climate, in both senses of the word. The winter period of Cassegrain observing with the traditional instrument suite was badly affected by weather, but the queue observing system meant that most of the highly rated projects were done nonetheless. On a more positive note, Tom Kerr became the

RCUH employee of the year for 2007, an award reflecting the huge effort he had put in over the preceding year.

Despite the successes in broadening the international use of UKIRT and the positive outcome of the Ward report, when PPAN reconsidered its rankings they rejected the Ward panel's recommendation regarding UKIRT, it remained a low priority. This was the final result of the Programmatic Review, and it led directly to a decision, in December 2008, to switch UKIRT to 100 % survey operations from early 2009 and to complete the UKIDSS work by mid-2012. The UKIRT Newsletter reported that while the commitment to UKIDSS completion was '*gratifying*' in the light of the severe financial pressures and the original intention to close it much earlier, the end of general purpose observing was a decision taken by the UKIRT board '*After much deliberation and with much regret*'. There was more than one reason to lament the change; a survey only telescope would be of less interest to potential partners, making finding new resources even more difficult. The long serving Cassegrain instruments were used for the last time on the 25th of January 2009. Ironically an upgrade to CGS4 had been completed in December, but this was no longer of much value since the instrument joined UIST and UFTI in limbo.

However, the dust had yet to settle. The much maligned STFC Programmatic Review in 2007–8 had prioritised the STFC support for operational optical-IR and radio astronomy facilities, but widespread dissatisfaction with the process caused the STFC management to promise another, longer term review, in 2009. This Ground-Based Facilities Review was chaired by Professor Michael Rowan-Robinson, and was charged with examining all STFC-operated optical, infrared, sub-millimetre and radio telescopes and relevant international facilities. Its aims included identifying the key science goals for ground-based astronomy at wavelengths ranging from the optical to radio in the next ten years and providing recommendations on the UK requirements and future strategy for a wide range of existing and planned telescopes.

The Panel's report was 50 pages long and covered a wide range of topics and recommendations, but of UKIRT it said '*There is a very strong case to support UKIRT to 2012 to complete the UKIDSS legacy surveys.* [...] *The case for continuation of UKIRT in 2012–2014 depends strongly on the completion of the proposed UKIRT Planet Finder (UPF) by 2012*' adding that '*The scientific case for UPF is very strong, but the Panel recognizes STFC's financial position may make it hard to deliver UPF in the period 2010–12 without substantial contributions from other sources*'.

## Planet Finder

The plan to build a planet finder for UKIRT grew out of a proposal originally made for an instrument to be mounted on one of the Gemini telescopes. The Precision Radial Velocity Spectrograph (PRVS) was intended to search for low mass (Earth-sized) planets using the same technique that was proving so successful in detecting large extra-solar planets at a number of telescopes around the world. The radial velocity method looks for tiny changes in the spectra of stars which arise when the gravitation tug of an orbiting planet causes its central star to wobble back and forth in a regular pattern. The resulting Doppler shift of the central star's light is most pronounced when the planet is massive and orbiting close to its parent star, so it can most easily find the "hot Jupiters" which dominated early planet discoveries. Small planets, the argument went, would be easier to find by searching lower mass stars, more likely to be affected by the tiny mass of an Earth-like world. However, unlike large hot stars which are best studied with visible light, the smaller K and M dwarf stars have many interesting spectral lines in the infrared region and are brightest at these wavelengths. So what was needed was an infrared equivalent of the high precision optical spectrographs (such as the HARPS instrument at ESO) which could be installed on a large telescope.

Thus was born the Gemini PRVS proposal, which was well developed and was on the point of being agreed when the financial crisis which accompanied the formation of STFC struck. Rumours of a UK withdrawal from Gemini were circulated, then denied, but the seeds of doubt regarding the UK's commitment to Gemini were sown and when there were funding cutbacks at Gemini, the PRVS project was abandoned. The cancellation of PRVS did however create an opportunity for Gary Davis who says '*I conceived UPF, the UKIRT Planet Finder, specifically in response to the low ranking of UKIRT in the Programmatic Review. When PRVS was cancelled I suggested we develop a proposal to build it for UKIRT. The idea was to set out an ambitious future for UKIRT in which it would have a unique, world-leading role in a sexy area of science, rather than relying on the excellence of the current science programme. This was a deliberate strategy*'. Discussions were held with Hugh Jones, the PRVS Project Scientist, and with Ian Robson, who was by now the Director of the UKATC from where the UPF work would have been led.

The proposed UPF would be installed in the basement of the UKIRT building, and fed by light directed down fibres from a module mounted on the WFCAM cryostat. To mount the fibres the top portion of WFCAM, which held the large corrective field lens, would be removed and replaced by a new module. This interchangeable module would make switching from one mode (surveying) to the other (planet hunting) fairly quick and easy since the bulk of the WFCAM could remain in place. Furthermore removing and replacing the top tower was a standard WFCAM operation during removal and re-installation, so was already well understood and well practiced. A survey of a large number of nearby stars with UPF would require about half of the telescope time over five years and could be fitted in around the ongoing UKIDSS and future surveys.

Since most of the design work was already done it was believed the UPF could be built quickly, in three years or so, and cheaply, for perhaps just 5 or 6 million pounds. However, getting approval for such an instrument would be a tortuous process. The first step was to submit a Statement of Interest to the STFC's committee for Particle Physics, Astronomy and Nuclear Physics. This was done by Hugh Jones in early 2009 and the concept was endorsed by the Committee. This was a very significant step forward and the next was to prepare a full proposal for another STFC committee, the Projects Peer Review Panel, in May. The approval process was expected to take several months, so a final verdict was not expected until late 2009, but nevertheless Gary Davis was able to write that '*It is gratifying that we have cleared the first hurdle*'.

However, although UPF was on its way, it would not reach the finish line. It was soon clear that STFC were not going to save the required £80 million resulting from the CSR settlement and approaching the third and final year, big cuts still had to be made. PPAN were instructed to develop a plan during the early autumn for presentation to the STFC's science board. A preliminary version was presented to STFC on the 19th and 20th of October at which time the members of PPAN were given some strategic advice and asked to develop a prioritised plan. At its next meeting on November the 30th and December the 1st the panel produced a ranked list of potential projects across the whole PPAN area, rating UPF and UKIRT in the alpha 4 category and noting that, within the budget envelope allowed, only projects in the alpha 1 and alpha 2 bands could be funded. The UPF proposal was all but dead and the final nail in the coffin was delivered on the 16th of December 2009 when STFC announced its plans for the next five years in a press conference. Despite the powerful scientific case, and the high rankings give the proposal by an independent review panel, there would be no UPF and, as a result, UKIRT was to close as soon as possible.

In his on-line blog on Sunday, 24 January 2010 Tom Kerr contrasted the views of Lord Rees, who was quoted on the BBC news web page, with the actions of the STFC. '*Lord Rees, the President of the Royal Society and Astronomer Royal, said such a discovery would be a moment which would change humanity. If it wasn't so depressing I would be laughing. The UK's STFC has decided not to fund the UKIRT Planet Finder (UPF) and as a consequence is withdraw funding from UKIRT. The UPF would have allowed the UK to detect these earth-sized planets and allow the country a chance to lead the world in this research*'.

It was ironic that while all the politicking had been going on, UKIRT continued to do world-beating science. In April 2009 UKIRT observed what turned out to be at the time the most distant object known in the Universe. A request to observe a gamma-ray burst detected by the SWIFT satellite had come at the beginning of a night when the weather was marginal with strong winds. Thor Wold and Tom Kerr were at the summit, but had not opened the dome because of the weather. However when they got the gamma-ray burst request, which requires immediate follow up to detect any optical or infrared afterglow from the burst, they decided to try. So they opened for about 20 minutes before the target set, took the data and then closed up again. The next day Nial Tanvir from Leicester University called back and was very excited that they had found something interesting. Tom and Thor had been the first to observe it and Tom says that just that feeling of being the first to see it was very exciting.

Soon after the gamma-ray burst observation UKIRT had passed its 30th anniversary and a workshop to celebrate these 30 years of achievement was held at the

ROE. Attended by upwards of 60 people it was a huge success, noting that UKIRT had evolved out of all recognition from its origins as cheap light bucket into a world-class facility that was still producing world class science and had the potential to keep on doing so for a number of years.

**Fig. 20.1** Some of the participants at the UKIRT 30th anniversary conference in Edinburgh (Photo Jason Cowan, UKATC)

Taking this opportunity to look back as well as forward, some of the senior staff at the meeting, recognised the contributions of Jim Ring, David Beattie and Sidney Arakaki, all of whom had by now died and others reflected on another more recent loss, technician Mark Horita had died on 11th August 2009 after 25 years at UKIRT. Although retired, Tim Hawarden was also present at the meeting where he gave a review of the upgrades programme in his usual dynamic and inimitable style. Sadly, it was to be Tim's last contribution to UKIRT for he died suddenly and unexpectedly on November 10th.

# Chapter 21
# Battle for Survival

The decision to cancel the UPF was a body blow for UKIRT, the more so because many people recognised the potential of the new instrument, but the deed was done and UKIRT was identified for closure on the 31st of March 2010, by then only a matter of a few months away. Gary Davis secured a stay of execution and the policy was changed to what was politely called a "managed withdrawal". So Gary set about trying to somehow save the situation, for example by finding a new owner and handing over the telescope to them. Despite what was undoubtedly a very bad hand, he did have one card left to play, UKIRT was only one of two telescopes operated by the JAC, and much of the support infrastructure, including the technical staff, was shared with the JCMT. Closing UKIRT would throw all of those costs onto the JCMT, making it more expensive to operate. The trick would be to find a way to cut UKIRT's costs so much that it would become more attractive to continue operating UKIRT than to close it. So he asked Andy Adamson and Tom Kerr to consider a number of possible scenarios, one of which was remote operation of the telescope from Hilo. Tom and Andy both quickly realised that remote operation was in fact a viable option and Tom recalls Gary being surprised that such a radical change in operational model would be possible in such a short time.

## Minimalist Mode

Thus was born what was to be called operating UKIRT in minimalist mode. This was a breathtakingly bold strategy to drastically reduce running costs by a factor of two, dropping the marginal cost of continuing to run UKIRT below the extra cost of operating JCMT alone. It would involve a streamlined science programme that no longer needed to be supported by visiting observers, cutting engineering support to the minimum and accepting a higher risk of losing time to technical problems. Its key lay in modifying the 30 year old telescope, designed in an analogue era, so it could be operated remotely from the Hilo office. The telescope and its building



J.K. Davies, *The Life Story of an Infrared Telescope*, Springer Praxis Books,
DOI 10.1007/978-3-319-23579-0_21

would be briefly inspected by JCMT staff at the beginning of the night to ensure that the dome could be opened safely but, busy with their own tasks in a building a kilometre away, they would be unable to intervene at UKIRT except in an emergency. Although the concept of remote operations had briefly arisen, and been rejected, during the financial crisis five years earlier minimalist mode looked to some like a fairly desperate gamble but Gary Davis disagrees. He describes it as '*A carefully measured response to a difficult set of constraints. It was extreme, certainly, but not desperate and not a gamble*'. The plan was approved by the UKIRT board in April 2010 and STFC committed itself to continue to operate the telescope until at least the 31st of March 2012. There was a certain irony to this turn of events. Some years earlier Ian Robson had pushed forward plans to have the JCMT operated remotely from the UKIRT building as way of making operational savings. This scheme would have been feasible since, unlike UKIRT, the JCMT had originally been designed to be operated more-or-less remotely. That plan never came to fruition and instead it was to be UKIRT that would become the remotely operated telescope dependent on its "younger sister".

Under the new scheme, scientific priority would be given to UKIDSS observations and a few of the campaign science projects. There would be no service observing and no open time for small projects, ending the need for a Time Allocation Committee and its associated overheads. All non-essential tasks would be cut. There would no newsletters, no public outreach, no development of new facilities or instruments and fewer staff astronomers. This plan would save on the costs of travelling to, and staying on, the mountain itself, but it would also allow staff numbers to be cut, 10 positions would be lost including Anna Lucas who had been providing administrative support for UKIRT since the year the telescope opened. Andy Adamson left, headed to a senior position with Gemini. Andy's departure was not a big surprise, he himself had identified about year before that when UKIRT entered minimalist mode, with no plans for future development, his own job as Associate Director would become redundant. With Andy gone, Tom Kerr was promoted to a new post called Head of UKIRT Operations.

There were other changes too, Brad Cavanagh, who had worked for eight years in the software group had already returned to Canada and software engineer Sam Hart, who had been at JAC for five years, returned to Britain when his US visa ran out. Frossie Economou, who had contributed so much to ORAC and the OMP and who had been working part-time since having a baby some years earlier departed for good. One of the resulting holes in the software group was plugged by survey scientist Luca Rizzi, who became a scientific programmer over the summer. Chris Davis, at UKIRT since the mid-1990s, went first to the JCMT and then left on a two year secondment to NASA headquarters in Washington before finding a job back in the UK in 2012. The staff reductions, plus savings in other costs such as requiring only one person working each night and the abolition of overnight stays at Hale Pohaku had the effect of reducing UKIRT's annual operational costs from $3.7 million to about $2 million.

The new operating mode would begin in December 2010, which meant that a lot had to be done in a very short time. A number of things were required before the switch could be made, most importantly making the operation of the telescope both technically robust and physically safe. The UKIRT team went through the list of actions required to run the telescope and to solve faults and tried to identify any which required direct human intervention, for example pressing a switch or turning a knob. Each of these was either modified for remote operation or alternative ways of getting the same effect (for example remotely rebooting computers) were identified and implemented.

According to Tom Kerr the hardest things about moving to this new model were to do with safety protocols. Since the telescope and instruments were already operated from a control room, and since network connections between Hilo and the summit were very good, he believes that they could have implemented the remote scheme almost immediately, but it would have been without proper safety systems in place. There was the need to set up protocols for various failures, such as loss of network connections or of power and to make sure that the telescope would be safe under these conditions. There was also the matter of physical security, with no-one in the dome at night it might it become a target for physical vandalism, perhaps by groups opposed to development on the summit or accidental damage from over-enthusiastic tourists trying to get into the dome. There was clearly a need for cameras, but there are very strict rules on changing the appearance of the buildings on Mauna Kea and sticking cameras willy-nilly around the outside was not an option. So a single infrared camera with a motion sensor was mounted to cover the UKIRT front door. Inside the dome no such cosmetic restrictions applied so cameras monitoring the dome, telescope and mirror covers were fitted. The following year further cameras were installed, along with microphones, allowing the physical security of the dome to be inspected from Hilo and removing the requirement for a visit by the JCMT observing team before operations could be started each night (although the requirement that the JCMT team are at the summit in case of emergency would remain).

Another obvious requirement was a sea level operations room, or UKIRT Remote Observation Centre (UROC). Ad-hoc tests had been done from the JAC computer terminal room but this was not a suitable long-term solution. There was a need to set up a sea level control room with which the TSS team were comfortable, something which mimicked the set up at the summit with the same screens in roughly the same places. The new remote control room was built up in an office near the JAC's rear door. A horseshoe shaped control console, with 4 large monitors controlled via a single keyboard, was built into an alcove. From here the UKIRT TSS could connect to the three computers in the UKIRT building that controlled the telescope, its data acquisition system and the data reduction pipeline. Another keyboard controlled another set of monitors, which fed back the weather data and the output from the CCTV cameras. To mitigate the difficulties of the remote operator trying to sleep during the day, a two bedroom apartment was leased on the seventh floor of the Hilo Lagoon Center, overlooking the hopefully peaceful Wailoa State Park in Hilo. In the new regime the TSS would visit the JAC in the late afternoon to discuss the night's programme and to run the requisite instrument calibrations. The TSS would then assume responsibility for the operation throughout the night, working alone in the Hilo office until morning.

**Fig. 21.1** JAC Director Gary Davis and TSS Jack Ehle in the new "UKIRT" control room, about 50 km from the telescope itself (Photo Tom Kerr)

The typical plan would be for the UKIRT TSS to run up the observing software and start taking initial array test data and calibrations about an hour before the JCMT team were due to leave Hale Pohaku for the summit. The JCMT team would visit UKIRT to inspect the area both inside and outside the dome and give the all clear to open the dome from sea level. They would then leave and have no more interaction with UKIRT that night other than communications with the Hilo operator or in the case of an emergency, for example if the UKIRT dome became inoperable. For safety reasons it was decided that UKIRT would close when the JCMT observers left the summit, so UKIRT's operating hours became dependent on the JCMT schedule. A further safety precaution was the writing of software to close the dome if either there was no action taken by the UKIRT TSS in an hour, if communications were lost between Hilo and the summit or if mains power was lost at the summit (the dome can still close using power from the battery powered emergency system).

## Remote Operations Begin

On the 13th December 2010, after a few trial runs of a few hours each, UKIRT was operated in a fully remote mode for the first time. With Omar Almaini, his research student Caterina Lani, and Tom Kerr at the summit, Tim Carroll ran the telescope, instrument and observing queue from Hilo with little or no intervention from those in the summit control room. This was repeated for one more night with little for the summit team to do except spend some time taking photos, drinking coffee and chatting about what UKIRT meant to them. Omar's view was one that many of his contemporaries could have echoed '*Most of my career over the last 15 years has been closely tied to this telescope. I've been out here perhaps a dozen times and I've grown very fond of the place. I've heard so many other astronomers with similar stories. I know of no other telescope that generates such affection in people. It really is a very special place*'.

On Wednesday the 15th of December the telescope was operated with an empty summit control room by Jack Ehle from Hilo. It was an emotional moment for both Omar and what was left of the UKIRT team. Tom Kerr admits that when he had started to pack up his gear to head down the morning before he '*Nearly lost it*'. One minute he felt fine the next he had '*Tears forming and just had to stop*'. Writing in his blog he noted that '*It seems many of our final visiting observers have felt the same way, one or two wanting to hug the telescope on their final night and the previous observer being cruelly denied from saying good-bye to the telescope by a winter storm. Even us hardened scientists can be emotional when it comes to inanimate objects! Although perhaps it's just because UKIRT has been responsible for making many people's careers in astronomy, including my own*'.

Omar seems to have felt much the same way, in his final observing report he wrote '*Apologies for the delay in writing this report, which I found quite difficult to write. I personally owe so much to UKIRT, so this was a very special run. It was an honour to witness the transition to remote observing while at the same time extremely sad to realise that we may be the final visiting observers. Hopefully some day soon we will have astronomers trundling up the mountain once again to use this wonderful telescope*'.

## A New Way of Working

Almost from the beginning the remote operations model worked extremely well, probably much better than even its proponents had expected in their most optimistic moments. Thor Wold, who had been working at UKIRT for 27 years by then, says that things '*turned out FAR better than I remotely expected. I thought I would really miss the summit - after all it is a rather awesome and beautiful place - but I really don't*'. Part of this surprise was almost certainly the relief from having to cope with the effects of altitude and the daily commute to summit. This alone made sea level working much less tiring and one TSS remarked that a 14 hour shift in Hilo seemed about the same as a 9 hour one at the summit. The arrangements for a quiet apartment in Hilo also worked well, allowing the TSS team to get some rest while the rest of the town went about its normal daytime business.

Worries that the TSS would find working alone either stressful or boring did not materialise. Tom Kerr had been concerned that they might find it hard to cope, with lots of new things for them to learn, but in practice he feels that it worked very well. The software indicates to the TSS if a particular observation is satisfactory or has to be re-done, and the queue scheduler software puts up the optimised observing plan for the night. Tom was also concerned that the TSS would miss the company of astronomers, but it turned out that this was not the case, they like to see old friends but they are really happy not having some-one else to look after. Thor Wold agrees. He discovered that '*I really like working alone. I like running this show by myself. No more sitting there* [*with visitors*] *thinking...go on, push the button, click that mouse, you can do it... and waiting for the visiting observer to absorb what is next*'. Of course the new scheme put even more responsibility onto the TSS if things did not go well, but once again they seemed to have relished the challenge with Thor remarking that '*When stuff breaks it is all the more your responsibility to get on it and get it going again*'.

In fact, the failure at rate at UKIRT remained very low, even though it would seem reasonable to suppose that it would get worse with no-one at the summit. For many years the target was to lose no more than 3 % of observing time to faults and this rate has been maintained despite the switch to remote operations. While some traditional night-time problems took longer to resolve than would have been usual if the summit had been occupied, the first telescope time lost to a fault due solely to remote operations did not occur until 14 months after start of this mode when the telescope computer, mistakenly thinking that communications with Hilo had been lost, closed the dome while observations were underway. UKIRT's observing efficiency, the fraction of clear time spent actually taking data, remained unchanged at about 90 %.

## Record Breakers of a Different Kind

The implementation of minimalist mode did more than just save UKIRT from immediate closure; it kept the telescope at the forefront of astronomical research. The year 2010 was very productive, with scientific papers based on data taken at UKIRT appearing at a record rate, making it the most productive year in its 31 year history and surpassing the record set only one year earlier. On the 10th of May 2011 it was announced that funding of the JAC would continue until at least the 31st of March 2013 in order to allow the JCMT to take advantage of its new flagship instrument SCUBA-2 and so UKIRT operations, now almost entirely focussed on the completion of the UKIDSS surveys, could continue until this date. Unsurprisingly 2011 proved another bumper year for UKIRT science with over 100 scientific papers published, a highlight being the discovery of a quasar with a redshift of greater than 6.4 which was found in the 3rd survey data release.

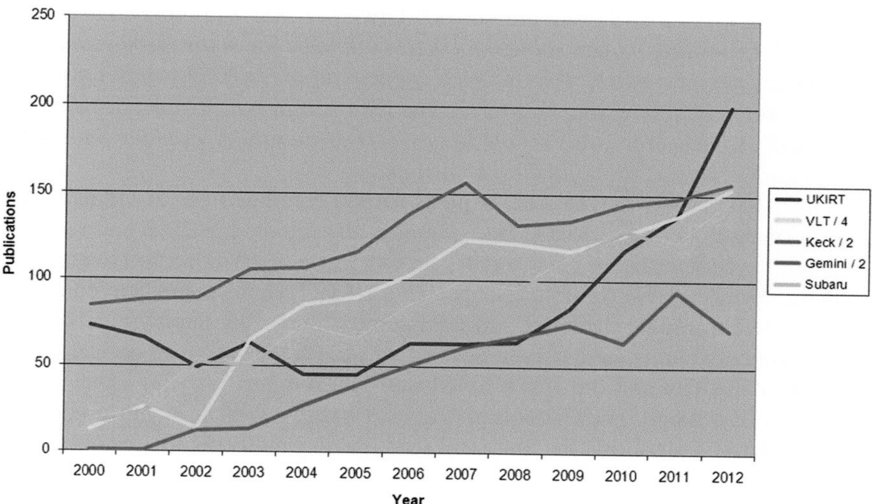


**Fig. 21.2** Papers based on UKIRT data have steadily increased in recent years as more and more UKIDSS data has become available. After allowing for the multiple telescopes which make up the VLT, Gemini and Keck Observatories, in 2012 UKIRT produced more papers per telescope than any major facility (Image JAC)

Gary Davis admits that the remote operations scheme had worked far better than anyone expected and it had made UKIRT, in terms of scientific papers produced per US dollar of running costs the most cost effective major telescope in the world. Indeed even the cost per paper qualification soon became irrelevant, by some measures at least, in 2012 UKIRT was the most productive telescope in the world.

The power of UKIRT for infrared surveys was re-emphasised in April 2012 when it was announced that the Korean Astronomy and Space Science Institute (KASI), had acquired 22 nights of UKIRT time in 2012. The KASI wished to carry out a programme to map 110 square degrees of the galactic plane using a bespoke filter sensitive to emission from $Fe^{+}$ ions which is a good tracer of dense, shock-excited gas and which would provide valuable insights into star formation and stellar feedback mechanisms. With the continuing interest from the Korean Centre for the Exploration of the Origin of the Universe from Seoul National University, Korean astronomers had access to 62 nights of observing time in 2012.

Meanwhile UKIDSS formally completed its observational programme in May 2012, seven years after starting observations in 2005, and largely as originally proposed in 2001. There remained a need for some minor filling of gaps, but by May the areas and depths were at about 80 % of the originally proposed size, and fairly precisely as set out in the revised plan of 2009. When the survey finished much of the data had already been made available to the astronomical community via a series of data releases, but further releases were made during 2012–2013 to complete the process. The final act was one more re-processing of all the data, using the lessons learned, to produce the final definitive legacy dataset. In summarising he UKIDSS programme Andy Lawrence said '*We expect that astronomers will be writing papers using the UKIDSS dataset for many more years. Like other major surveys, much of the science coming out of UKIDSS would have been hard to anticipate fifteen years ago when the idea was first developed. However, it has clearly achieved the goals set out in the 2001 proposal, for example finding the coolest known star, and the furthest known quasar*'.

With UKIDSS drawing to a close plans for the continued use of UKIRT were already being drawn up in 2011. One proposal, first mooted in 2006 and revisited from time to time in later years, was for an extension of UKIDSS to cover all the northern sky accessible to the telescope. This UKIRT Hemisphere Survey (UHS) would produce a huge database and complement other sky surveys including a similar southern hemisphere survey by the 4 m VISTA telescope, an optical survey called PanSTARRs and the space based mid-infrared survey being done by the NASA WISE mission. The UHS observing plan was to cover the sky in single filter (initially the J band) during the first year, then obtain equally complete coverage in another filter (K) the following year. Assuming that UKIRT operations continued, the J and K coverage would repeat in years three and four. A few targeted projects covering more localised areas were included to maximise the scientific return and allow some observing flexibility. The plan was put to the UKIRT director Gary Davis in the spring of 2011 and approved, initially for one year of operations, on the 3rd of June 2011. The first observations were taken on the 19th May of 2012.

# Chapter 22
# The Axe Falls

Despite the high productivity and low operational costs of UKIRT in its remote operations mode the continuing pressure on the UK astronomy budget resulted in yet another review of the two remaining British overseas observatories. This was begun in late 2011 and concluded in the spring of 2012. Even as UKIRT's 2011 record breaking productivity was being announced, and the new UHS programme was beginning, a fateful decision was being made at the highest level of UK astronomy. On the 31st of May 2012 it was announced that STFC Council had considered the future of the two sites in La Palma and Hawaii and, acting on advice and recommendations from its Science Board, had agreed to:

- extend operation of JCMT to end September 2014, to allow for completion of the agreed science programme for the SCUBA-2 instrument on the JCMT;
- cease STFC support for the operation of UKIRT from end September 2013, a year after the completion of its current survey programme;

The announcement went on to say '*We are pleased to be able to extend the operation of the Hawaii telescopes for at least another year to enable further excellent research to be conducted* [. . .] *However, we must now also commence negotiations with the University of Hawaii as the leaseholder of the Mauna Kea sites, and with other potential operators of each of the Hawaii telescopes. If a suitable alternate operator is not identified for either Hawaiian telescope, STFC will decommission that telescope and restore the site as required by the lease*'.

The news was a bitter blow to the UKIRT community, the more so because it was made when the UKIDSS consortium were just about to start a two-day scientific meeting to discuss the results still emerging from the survey data and the beginning of the new UHS programme. According to one participant it '*turned what should have been a celebration into a wake*'. Tom Kerr's entry to his blog that day was headlined simply "Devastated" and after echoing the STFC announcement went on to express incredulity at the decision to close UKIRT a year before the termination of JCMT operations. Tom expressed the feeling of many who could see no logic in the decision saying. '*The reason for ending support for UKIRT a year*


J.K. Davies, *The Life Story of an Infrared Telescope*, Springer Praxis Books,
DOI 10.1007/978-3-319-23579-0_22

*earlier than JCMT is unclear since there are no savings to be made here, UKIRT essentially comes for free as long as the JCMT is supported. The opportunity to take a year's worth of free data using one of the most scientifically productive telescopes on the planet is being thrown away*'.

The UKIRT board issued a more formal statement, re-iterating this view. After noting the financial pressures the statement went on to say '*We are however, very disappointed that STFC funding for UKIRT Operations will cease from 30 September 2013 and do not understand why the opportunity to continue scientific operations for another year has been rejected, particularly as the operations costs that would fall on STFC are very low. (Through contributions from international customers and the shared operation with JCMT, the additional funding needed to operate UKIRT to Sept 2014 is calculated to be less than £100,000). UKIRT is a world-leading facility; WFCAM on UKIRT remains the most sensitive wide-field IR camera in the world in the critical K-band and provides unique infrared survey opportunities for a large range of programmes including complementary data to some of the highest priority current and planned astronomy programmes*'.

The Royal Astronomical Society reaction was limited to an expression of '*deep regret at the decision of the Science and Technology Facilities Council (STFC) to end support for two major astronomical telescopes*' although it allowed itself to welcome the continued support for the telescopes in La Palma, which had survived the axe for the moment. However David Southwood, president of the society, agreed that the Hawaiian telescopes were doing '*front-rank*' work and described their closure as '*a sad day for British astronomy*' adding that he had resisted pressure from some astronomers to "jump up and down" on the STFC because he recognised that its hands were tied by the constraints on its budget.

## The Next Chapter

The STFC announcement, while deeply sad to many of UKIRT's users past and present, was not the end of the story. Prof John Womersley, the STFC Chief Executive, made it clear that STFC would be happy to hand over UKIRT as a going concern to anyone who might wish to take it on. This new operator might be a foreign scientific agency or a consortium of Universities, either from the UK or elsewhere. He indicated that the cost to the new operator was likely to be small, perhaps a nominal amount such as £1 or $1, provided that the new operator took over from the UK the responsibility to eventually decommission the site. To sweeten such a deal any new operator was expected to be offered a dowry, in the form of the estimated decommissioning costs, which STFC had already of necessity set aside in case they were needed.

So, in 2012 the JAC published on its website a notice drawing to the attention of the world that the UKIRT was available to any credible taker. This announcement said in part that '*The observatory, its instrumentation and its support equipment are therefore being offered to the global astronomical community through this Announcement of Opportunity. There are no preconceptions or constraints: we welcome parties wishing to take over the entire observatory, parties interested in being minor partners, and any other permutation. We are willing to consider any and all possibilities. Details of the facilities being made available and the process for registering your interest are all described in the prospectus. This is the first time that a productive, world-leading observatory in the 4-m class has been offered to the global community. We invite you to consider this unprecedented opportunity*'.

The linked prospectus was a 28 page colour booklet, not unlike the sort of brochure given away in car showrooms or housing developments, which set out the details of the telescope, the instruments (including Michelle), the site, the estimated running costs and so forth. It concluded with a summary of scientific highlights and an invitation for potential new operators to make an expression of interest by the 30th of November 2012.

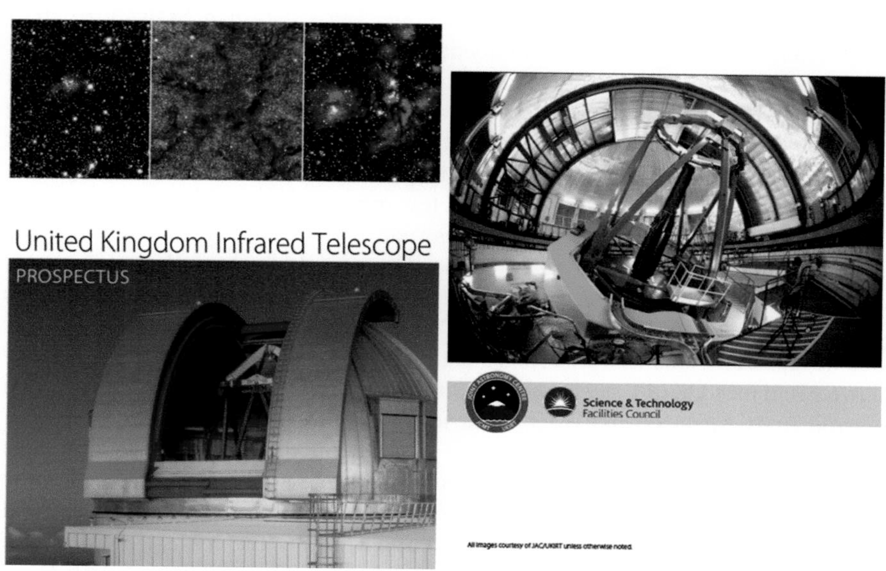


**Fig. 22.1** The UKIRT Prospectus, front and back pages, issued in the Autumn of 2012 (Photo STFC)

This was uncharted territory, it is hard to think of any major national telescope anywhere in the world that has been handed over in full working order to a new owner. It is even rarer to find one which has been decommissioned and then razed to the ground, one path-finding step most of the UKIRT community would have hated to see their telescope take. However, this most pessimistic scenario never seemed very likely and while details were scarce, it soon became generally known that several possible options were under discussion. The situation was clarified in the summer of 2013 when STFC made an announcement that from a dozen or so expressions of interest two groups had been invited to develop more detailed proposals. In parallel to these developments it was also announced that the University of Hawaii had agreed to assume legal ownership of UKIRT once STFC funded operations ceased and thus to take over responsibility for the site on Mauna Kea. This would allow the British to terminate their existing sub-lease and release them from the potentially expensive burden of decommissioning the site should no future owner be found. However the process of legally transferring ownership to the University, and allowing the University to establish a partnership with a new operator was something which was bound to take time, and in July STFC announced that UKIRT operations would continue until the end of the year.

By the Autumn of 2013 no announcement had been made as to who the new operators of the telescope would be, but the situation was starting to get rather farcical as the outcome was being widely rumoured amongst those working on the summit and so had already leaked to astronomers in Europe. The staff of the JAC was unable to confirm or deny these rumours since they had been given strict instructions that this issue was confidential and, since they hoped to retain their positions within the new regime, they were hardly likely to go on the record against this policy. However, in fairness, it was no longer really their secret; the impending transfer of responsibility to the University of Hawaii had placed the onus on that organisation to break the news. In the meantime UKIRT continued to operate, trying to complete the UHS survey and a few other projects which had been badly affected by bad weather over the previous year.

The end of September, originally the date on which the UK would finally withdraw from UKIRT, did mark another milestone when Thor Wold, UKIRT's longest serving TSS, finally retired. Thor's last shift was on the 22nd of September and he received a small surprise, and a bottle of Glenlivet whisky, from a handful of current and past UKIRT staff. The party included Tom Kerr, Dolores Walther, Tim Carroll Andy Adamson and Watson Varricatt, who dropped in on him at the start of his final shift in the remote operations room at the JAC. The night itself was uneventful with low winds, clear skies and good seeing, although Thor admitted to the experience of his final shift being '*surreal*'. A more public event, in the form of a farewell party for both him and Jack Ehle, who was also leaving at the end of September after a number of years, was held at the Onekahakaha Beach Park. Amongst the other tributes to Thor was a document containing goodwill messages from some of the many astronomers with whom he had worked over the 28 years he had been in post. Two new staff were recruited and trained to replace Thor and Jack. They were Sam Begnini and Erik Moore who were both physics graduates of the University of Hawaii in Hilo with experience of working at the UH 88 in. telescope.

The 31st of December 2013 came and went, with no official announcement as to what was happening although by then it was an open secret that Lockheed Martin and the University of Arizona were going to be the new operators. So UKIRT operations continued although behind the scenes the funding was coming from NASA (via Lockheed) and not from STFC. This was expected to be a short term transitional arrangement until the legal ownership could be transferred to the

University of Hawaii but the transfer took longer than expected. There was a plan to handover the telescope in March 2014, and a ceremony was scheduled although this never took place due to continuing delays over the details. While the "What's new" pages of the University of Arizona's Steward Observatory had a link to UKIRT information dated March 26th, the lid came off in April when the University of Hawaii held a meeting which was open to the public and the local media picked up on the story. In articles published in the fourth week of April the Hawaii Tribune-Herald and Honolulu Star Advertiser (and of course many astronomy blogs) indicated that UKIRT director Gary Davis had confirmed that negotiations were underway between the University of Hawaii, the University of Arizona and the Lockheed Martin Corporation. Gary was quoted as saying '*we are transferring the telescope to the University of Hawaii which is entering into an agreement with the University of Arizona*'.

Richard Green, assistant director of the University of Arizona's Steward Observatory, indicated that he would become UKIRT's director when the agreements were finalized and that Lockheed's scientific research branch will use the telescope to conduct research that, as the newspapers put it, will eventually be published. Green said '*NASA has provided funding for UKIRT and that money will flow through a scientific research branch at Lockheed and Lockheed will partner with the University of Arizona for research. The telescope will belong to the University of Hawaii. That's what's under discussion and what's being worked out as we speak*'. Rather than a lease he described the new situation as '*a scientific cooperation agreement with UH*'.

According to the newspaper story NASA's interest in UKIRT involves protecting working satellites in space and studying properties of satellite material that is in orbit around Earth, although the two universities also have other potential projects, including collecting data for studying properties of asteroids that might come near Earth and studying properties of some of the earliest, most distant galaxies in the Universe. When asked why this was taking so long Davis and Green said they were waiting to complete legal agreements before making the transfer public and that they were hoping to wrap it up in the next few months. So by the end of 2013 the UK had essentially pulled out from UKIRT, although there was still some scientific involvement in the UKIRT Hemisphere Survey which continued as part of the Hawaii-Arizona astronomy programme. This co-operation was leveraged by the involvement of UK based groups in processing the data from the survey and even as UKIRT was being handed over discussions on a continuing, if small, UK involvement as part of a shared science programme continued.

Meanwhile the new operators were clearly intending to make the most of their telescope. In October 2014 two of the old Cassegrain instruments, UFTI and UIST, were brought back into service by astronomers working physically at the summit. Over the following nights these instruments were commissioned for remote operations from Hilo, something which had never been done during the minimalist mode sky surveys of the preceding few years. On the second of these nights the whole time was spent taking science data. Commissioning of the modes to take orbital debris observations followed shortly after. Enquiries were also made to the

UKATC about upgrades to the WFCAM and possible re-commissioning of Michelle.

So once again UKIRT survived by being re-invented. The observatory had evolved from a cheap "do it yourself" flux collector in the 1980s to a world class, optical quality, general purpose telescope with active optics in the 1990s to a wide-field survey instrument in the twenty-first Century. Its nightly operation went from teams of six armed with pocket calculators and digital voltmeters to a computer driven, queue scheduled discovery machine operated by a single person sitting in room 50 km away. In 2014 a new chapter opened, who knows how the story will end?[1]

[1] On 21 October 2015 a web post by the University of Hawaii, Hilo announced *that 'The University of Hawaii has identified the third observatory to be decommissioned and removed from the summit of Mauna Kea, advancing the implementation of the Mauna Kea Comprehensive Management Plan. The third observatory is the UKIRT Observatory, formerly known as the United Kingdom Infrared Telescope. Detailed planning for the removal of the UKIRT observatory and restoration of the site will begin sometime after the decommissioning processes for the Caltech and Hoku Kea* [UH 24 inch] *observatories and will be completed in accordance with the governor's plan. No new observatories will be built on the three sites.'* No date was given but decommissioning of the Holu Kea telescope is expected to be complete by 2018. It thus seems possible that UKIRT will just reach its 40$^{th}$ birthday in 2019.

# Chapter 23
# Epilogue

The STFC finally transferred ownership of UKIRT to the University of Hawaii at the end of October 2014. With the cessation of UK-funded operations the existing sublease for UKIRT was terminated. The facility and responsibility for the site, including the telescope, all instruments, associated equipment and software was transferred to the University of Hawaii who, as was already widely known, had negotiated an agreement with the University of Arizona and the Lockheed Martin Space Technology Advanced Research and Development Laboratories. The few remaining UKIRT staff had their existing contracts ended, but most of them remained employed after moving to new contracts under the responsibility of the new operators.

There was a celebratory event in Hilo on the 28th of October 2014, at the public 'Imiloa Astronomy Center which lies just across the street from the Joint Astronomy Centre. The day had begun with a visit to the summit for some of the VIPs some of whom had never visited the telescope they were transferring. Long time UKIRT user, and last chairman of the UKIRT board, Pat Roche joined the tour and recalls noticing '*how clean and tidy everything looked*'.

The main event was held in the evening. There was a reception at which speeches were given by representatives of STFC, the University of Hawaii, the University of Arizona and Lockheed Martin. Tom Geballe could not be present but he sent a message in which he said '*This versatile telescope has benefited from an ever-changing team of diverse, dedicated, and imaginative individuals, which made UKIRT a success for 35 years and which in its latest incarnation will surely continue to do so. UKIRT vastly exceeded the expectations of those farsighted individuals who conceived of it in the 1960s and early 1970s as well as those who brought it into initial operation. Now, as in many instances in the past, UKIRT is breaking new ground and embarking on a path that very few, if any of us at all, could have foreseen even a few short years earlier. I salute all of those who participated in UKIRT's rich past and send my congratulations and best wishes to the team that is traveling this new path for a large infrared telescope. As a*



J.K. Davies, *The Life Story of an Infrared Telescope*, Springer Praxis Books,
DOI 10.1007/978-3-319-23579-0_23

*perennial UKIRT fan, I look forward to learning more about the telescope's enhanced capabilities and its future findings*'.

The reception was followed by a public talk given by Andy Lawrence in the 'Imiloa planetarium, a show into which he had incorporated data from the UKIDSS sky survey in which he had played such an important part. The event continued with a meal and after dinner speeches by Pat Roche and Richard Ellis. Participants received a few small gifts, including a wooden coaster still bearing the date of the event originally planned for the 27th of March.

In the press note published on the STFC web-site Gary Davis was quoted as saying '*UKIRT has been a fabulous success story for British astronomy over its 35-year lifetime. I have never known a machine that inspires such affection amongst its users. Over the past decade, the UKIRT Infrared Deep Sky Survey (UKIDSS) has opened up new frontiers in infrared astronomy, and as a consequence UKIRT has been the most productive telescope on the planet for the past two years. It has been my honour to be the Director of this remarkable observatory*'.

UH President David Lassner said '*We are pleased to steward the UKIRT, a telescope that has made remarkable discoveries supporting the advancement of astronomical science. It is fitting to add it to our world-class portfolio of research assets, as the UKIRT has pioneered many operational innovations, including flexible scheduling and the provision of data reduction pipelines*'. Dr Guenther Hasinger, Director of the UH Institute for Astronomy, added that '*UKIRT will continue to produce top quality astronomical research. With a capable new operator and state-of-the-art instrumentation, UKIRT can continue to be a world leader in infrared astronomy for at least 10 more years*'.

The same release quoted Pat Roche as saying '*Astronomers using UKIRT have made many world-leading discoveries, including the detection and characterization of the weak emission from brown dwarfs to the identification of the most distant quasar known. UKIRT's innovative instruments have played a key role in the development of the field of infrared astronomy, with a rich stream of astronomical results supporting research programmes and student training at universities throughout the UK and beyond. The telescope remains a very powerful instrument at the peak of its performance, and I am confident that it will continue to produce exciting results under the new operational arrangements*'.

Away from the formalities the general feeling amongst several of the key players in the story of this remarkable telescope seems to have been that while few members of the UKIRT community believed this was the right thing to do, it was good to see that the telescope had a future, and that the new operators wanted to make the best use of it. Two of them used the same expression, it was a *bitter-sweet* moment.

# Index




J.K. Davies, *The Life Story of an Infrared Telescope*, Springer Praxis Books,
DOI 10.1007/978-3-319-23579-0

**D**



**E**



**F**

Printed in the United States
By Bookmasters